Lecture Notes in Computer Science　　9500

Commenced Publication in 1973
Founding and Former Series Editors:
Gerhard Goos, Juris Hartmanis, and Jan van Leeuwen

More information about this series at http://www.springer.com/series/7409

Dmitry Mouromtsev · Mathieu d'Aquin (Eds.)

Open Data
for Education

Linked, Shared, and Reusable Data
for Teaching and Learning

Springer

Editors
Dmitry Mouromtsev
ITMO University
St. Petersburg
Russia

Mathieu d'Aquin
Knowledge Media Institute
Milton Keynes
UK

ISSN 0302-9743　　　　　　ISSN 1611-3349　(electronic)
Lecture Notes in Computer Science
ISBN 978-3-319-30492-2　　　ISBN 978-3-319-30493-9　(eBook)
DOI 10.1007/978-3-319-30493-9

Library of Congress Control Number: 2016933113

LNCS Sublibrary: SL3 – Information Systems and Applications, incl. Internet/Web, and HCI

Printed on acid-free paper

This Springer imprint is published by Springer Nature
The registered company is Springer International Publishing AG Switzerland

Preface

The amount of open data, including especially linked open data, is constantly increasing in many domains, especially in the public sector. A great number of private and public organizations, institutions, and companies open their data and are interested in efficient solutions for sharing and reuse of published datasets. Obvious benefits come with opening data for end-users, organizations, and developers, by making it easier to find, obtain, and use data independently of their origin, the systems used to produce them, or the applications for which they are intended. This directly connects with the way the areas of learning, teaching, and education are evolving. Indeed, the activity of learning is changing very rapidly, especially through the Web, data, and open technologies. Distance learning is becoming more common, based on openly available educational resources on the Web and the recently appeared massive open online courses (MOOC) both in public higher education institutions and private training centers and organizations.

The primary goal of open data in education is therefore to support these changes through new methodologies and technologies that support the sharing and distribution of information about teaching and the subjects of learning. On the practical side, it serves various purposes such as to help teachers to find and create reusable educational materials, to assist students and families in their educational decisions throughout their life, to improve management systems and many others. For this reason the section of educational open data on the Web has expanded with information about courses and educational materials that can be accessed by tools and applications as well as, social and collaborative resources, thus shaping new architectures of open education. The past few years have demonstrated the growing interest in the topic of educational open data and the growth of the community. During five successful editions of the LILE (Linked Learning) workshops, keynotes, paper sessions, and panel discussions have shown the state of the art and progress in practical work with open data in education. A number of initiatives were started including community platforms (such as LinkedUniversities. org), the W3C Open Linked Education Community Group[1], and activities within the Open Knowledge[2] and the VIVO platform[3], to name just a few.

The goal of this book is therefore to act as a snapshot of current activities, and to share and disseminate the growing collective experience on open and linked data in education. In this volume we bring together research results, studies, and practical endeavors from initiatives spread across several countries around the world. These initiatives are laying the foundations of open and linked data in the education movement, and they are leading the way through innovative applications.

---

[1] https://www.w3.org/community/opened/

[2] https://okfn.org/

[3] http://www.vivoweb.org

The chapters are selected from extended versions of papers presented at an Open Data in Education Seminar[4] and the LILE workshops during 2014–2015[5,6]. They have been chosen to represent the diversity of practices and experiences that exist in the domain, from the researchers, developers, and community leaders who are pioneering the use of open and linked data in education.

In the first part of this book, two chapters provide different perspectives on the current state of the use of linked and open data in education, including the use of technology and the topics that are being covered.

The second part is to be considered the core of this book as it focuses on the specific, practical applications that are being put in place to exploit open and linked data in education today. In these four chapters, applications are presented ranging from the set-up of open data platforms in educational institutions, to supporting specific learning activities through the use of online, open data.

Finally, a key element of the evolving world of open data is to ensure the skills and ability to use such data are there. We therefore focus in the three last chapters of this book on the other side of open and linked data in education: on teaching the technology and practices so they can be widely applied, and on the community of practitioners pushing these practices forward.

We assume the readers of this book are reasonably familiar with modern educational technologies and Web standards (including basics of the Semantic Web). The chapters will be of interest, to varying extents, to academic heads and managers; educators, teachers, and tutors, and start-ups in education; library staff; postgraduates; technology researchers and professionals; as well as students and learners who are keen to better understand how the technologies of the Web and linked data can be applied to support progress in learning and education.

We acknowledge all the contributors and those who spent time on reviewing chapters and making critical comments and fruitful discussions. First of all we want to thank the members of numerous projects that have supported the development of the works presented in this book, including in particular the LUCERO project, the LinkedUp support action, the VIVO project, and some others. We also thank the funders of these projects, as well as our universities and organizations, especially the Open University and ITMO University that provided the environment for such projects to develop. We also want to thank all the members of the various communities dedicated to making open data in education a reality, including the W3C Open and Linked Education community group, the Open Knowledge Open Education Group, LinkedUniversities.org, and LinkedEducation.org. Finally, we thank our families, friends, and colleagues for their support and positive encouragement.

January 2016   Dmitry Mouromtsev
                                                        Mathieu d'Aquin

---

[4] https://linkededucation.wordpress.com/events/open-data-in-education-seminar-st-petersburg/

[5] https://linkededucation.wordpress.com/events/lile2014/

[6] https://lile2015.wordpress.com/

Contents

State of Open and Linked Data for Education

On the Use of Linked Open Data in Education: Current and Future Practices

Mathieu d'Aquin$^{(\boxtimes)}$

Knowledge Media Institute, The Open University,
Walton Hall, Milton Keynes, UK
mathieu.daquin@open.ac.uk

Abstract. Education has often been a keen adopter of new information and communication technologies. This is not surprising given that education is all about informing and communicating. Traditionally, educational institutions produce large volumes of data, much of which is publicly available, either because it is useful to communicate (e.g. the course catalogue) or because of external policies (e.g. reports to funding bodies). Considering the distribution and variety of providers (universities, schools, governments), topics (disciplines and types of educational data) and users (students, teachers, parents), education therefore represents a perfect use case for Linked Open Data. In this chapter, we look at the growing practices in using Linked Open Data in education, and how this trend is opening up opportunities for new services and new scenarios.

Keywords: Linked data · Semantic web · Education · Learning

1 Why Using Linked Data in Education

Traditionally, educational institutions produce large volumes of data, much of which is publicly available, either because it is useful to communicate (e.g., the course catalogue) or because of external policies (e.g., reports to funding bodies). In this context, open data has an important role to play. Implementing open data through Linked Data technologies can be summarized as using the web both as a channel to access data (through URIs supporting the delivery of structured information) and as a platform for the representation and integration of data (through creating a graph of links between these data URIs). Considering the distribution and variety of providers (universities, schools, governments), topics (disciplines and types of educational data) and users (students, teachers, parents), education also represents a perfect use case for Linked Open Data [7].

Indeed, the basic idea of Linked Data [9] is to use the architecture of the Web to share, distribute and interconnect data from various origins into a common, online environment. It is based on the basic principle that raw data objects are identified and accessible, similarly to webpages, through Web addresses (URIs), that deliver the information in a structured, processable and linkable way.

© Springer International Publishing Switzerland 2016
D. Mouromtsev and M. d'Aquin (Eds.): Open Data for Education, LNCS 9500, pp. 3–15, 2016.
DOI: 10.1007/978-3-319-30493-9_1

This approach has been very successful in the last few years, especially as a base method for the publication of open data on the Web. Linked Data has been adopted by government agencies in several countries (prominently, in the UK and the US) for transparency and public information purposes, by cultural heritage institutions such as libraries and museum to provide more processable and integrated information about their collections (see the Europeana project[1] for example, or the British Museum Collection[2]), by companies in publishing (for example at Nature, or Elsevier), broadcasting (for example at the BBC), or retail (for example at BestBuy). As we will see later in this chapter, there is a growing trend in the use of Linked Data specifically for education, with universities in particular making their public information (academic programmes, research outputs, facilities, etc.) available as linked data on the Web (see for example LinkedUniversities.org).

2   Linked Data - In More Details

The foundation of the Web is that it is a network of documents connected by hyperlinks. Each document is identified by a Web address, a URI, and might represent a document which content is encoded using a standard, universally readable format (most commonly HTML). The foundation of Linked Data is that data objects on the Web are identified, similarly to documents, by URIs. The representation of the data – i.e. the information associated with a data object – is then represented by Web links, which can themselves be characterised by URIs. This makes it possible to represent information in such a way that it is materialised as a graph, where nodes are URIs or literal data values (strings, numbers) and the edges are links between them.

For example, a university like The Open University[3] publishes information about the courses it offers through its website, as well as using linked data [3]. It achieves that through assigning to every course a dedicated URI that acts both as an identifier for the course on the Web, and as a way to address information about this course. For example, http://data.open.ac.uk/course/aa100 is the URI for the course with code AA100, which is an undergraduate (level 1) course in Arts and Humanities, entitled "The arts past and present". Through the links between this URI and others, information about this course is being represented regarding the topics and description of the course, where it is available, how it is assessed, what course material and open educational resources relate to it, etc. (see Fig. 1).

While most of the other data objects it relates to are also identified by URIs within the domain of the Open University, it is important to remark here that it links to other data sources, such as the UK government's information about The Open University or information provided by the Geonames platform about

---

[1] http://www.europeana.eu/.

[2] http://collection.britishmuseum.org/.

[3] http://www.open.ac.uk.

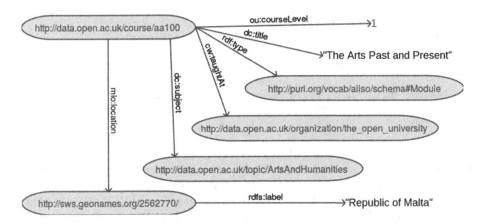

Fig. 1. Extract of the Linked Data (RDF) representation of the course AA100 "The Arts Past and Present" at The Open University (from data.open.ac.uk).

the countries in which the course is available. This demonstrates how, from these basic principles, information originating from widely different systems and sources can be seamlessly integrated.

Following the base principles described above, the most basic technology employed to implement linked data is a web-enabled, graph-based data representation language: RDF (Resource Description Framework [10]). RDF is in principle related to XML, but dedicated to the representation of graphs where nodes are URIs or literal values, and edges are links labelled by URIs. It has different syntaxes, including an XML-based one, but also others based on listing the triples [subject, predicate, objects] forming the links in the data.

Another important component of the technological stack for linked data is the one of vocabularies. Indeed, it is important that data should be shareable and reusable in a common way across sources and systems. To address that, languages such as RDF-Schema [2] and OWL (the Web Ontology Language [11]) allow one to define the types of objects that can be encountered in the data (the classes, e.g. Course, Person, Country, etc.), as well as the types of relationships that connect these types of objects (the properties, e.g. location, title, employer, author, etc.).

Finally, another important element of linked data is the way in which, still relying uniquely on the basic mechanisms of the Web, the data can be consumed. As we already mentioned, URIs on linked data can be requested to obtain RDF (most often in its XML syntax). When more flexibility is required, many of the existing linked data sources offer data endpoint using the standard querying language and protocol for RDF/Linked Data: SPARQL [12]. Briefly, SPARQL is both a query language made explicitly to fit the graph data model of RDF, and a Web protocol dictating the way in which a SPARQL endpoint should be accessed and queried on the Web. For example, the query:

```
select distinct ?course ?title where {
    ?course a <http://purl.org/vocab/aiiso/schema#Module>.
    ?course <http://purl.org/dc/elements/1.1/title> ?title.
    ?course <http://data.open.ac.uk/saou/ontology#isAvailableIn>
                        <http://sws.geonames.org/2328926/>.
    ?course <http://purl.org/dc/elements/1.1/subject>
                        <http://data.open.ac.uk/topic/computing
} limit 200
```

returns the courses (URIs and titles) in computing and IT offered by the Open University and that are available in Nigeria, when executed on the Open Universitys SPARQL endpoint[4]. Accessing such a SPARQL endpoint does not require any specific API or library, but is achieved using standard HTTP requests. The query above can therefore also be shared via a standard Web link[5].

3   The Adoption of Linked Data in Education

As described above, the elementary principle of the Linked Data of using the Web as a data modelling and access mechanism makes it effective to share and connect information from various sources. This is a property that many institutions have already started to exploit, and that is well aligned with the objective of many educational initiatives, especially related to open education: To disseminate knowledge resources and enable learning in a connected and global way. In this section, we therefore review the current adoption of these principles and technologies in the area of education, to understand how much this has already happened, and conclude in the next section with views on the next steps and the future of education with open, linked data.

We start our analysis with the LinkedUp project[6]. Indeed LinkedUp was a European project with the explicit objective to push forward the adoption of Web Data in Education. To support achieving this goal, the project developed a catalogue of education-related Linked Data sources that has grown to several dozen datasets in the last couple of years. Our methodology therefore relies on analysing the content of the LinkedUp Catalogue of Educational Datasets in order to understand the way in which Linked Data has been applied for education already, and what we can expect to happen in the future in this area.

---

[4] http://data.open.ac.uk/query.

[5] http://data.open.ac.uk/sparql?query=select%20distinct%20%3Fcourse%20%3Ftitle %20where%20%7B%3Fcourse%20a%20%3Chttp%3A%2F%2Fpurl.org%2Fvocab%2 Faiiso%2Fschema%23Module%3E.%20%3Fcourse%20%3Chttp%3A%2F%2Fpurl. org%2Fdc%2Felements%2F1.1%2Ftitle%3E%20%3Ftitle.%20%3Fcourse%20%3 Chttp%3A%2F%2Fdata.open.ac.uk%2Fsaou%2Fontology%23isAvailableIn%3E%20 %3Chttp%3A%2F%2Fsws.geonames.org%2F2328926%2F%3E.%20%3Fcourse%20% 3Chttp%3A%2F%2Fpurl.org%2Fdc%2Felements%2F1.1%2Fsubject%3E%20%3 Chttp%3A%2F%2Fdata.open.ac.uk%2Ftopic%2Fcomputing%2526it%3E%7D%20 limit%20200.

[6] http://linkedup-project.eu.

3.1 The LinkedUp Project and the LinkedUp Catalogue of Educational Datasets

The LinkedUp Project (Linking Web data for education [8]) was an EU FP7 Coordination and Support Action running from November 2012 to November 2014 which looked at issues around open data in education, with the aim of pushing forward the exploitation of the vast amounts of public, open data available on the Web. The project comprised six pan-European consortium partners led by the L3S Research Center of the Gottfried Wilhelm Leibniz Universitt Hannover and consisting of the Open University UK, the Open Knowledge Foundation, Elsevier, the Open Universiteit Nederland and eXact learning LCMS. The project also had a number of associated partners with an interest in the project including the Commonwealth of Learning, Canada, and the Department of Informatics, PUC-Rio, Brazil.

To aid awareness and use of open and linked data in education, the project created and has continuously maintained a catalogue and repository of data relevant and useful to education scenarios. The goal of the LinkedUp Dataset Catalog (or Linked Education Cloud[7]) is to collect and make available, ideally in an easily usable form, all sorts of data sources of relevance to education, providing a shared, evolving resource for the community interested in Web data for education (see Fig. 2). During the project, the technical team has enabled and encouraged content- and data-providers to contribute new material to the LinkedUp Dataset Catalog through a series of hands-on workshops and the promotion of community documentation on LinkedUp tools, workflows and lessons learned.

Datahub.io is probably the most popular registry of global catalogues of datasets and forms the heart of the Linked Open Data cloud. In the interest of integrating with other ongoing open data effort, rather than developing in isolation, the LinkedUp Data Catalog has been created as part of Datahub.io. It takes the form of a community group in which any dataset can be included, provided that it is relevant, and the datasets in this group are also visible globally on the Datahub.io portal. Every dataset is described with a set of basic metadata and assigned resources. This makes it possible to search for datasets and employ faceted browsing of the results both globally or specifically in the Linked Education Cloud. For example, one could search for the word 'university' in the Linked Education Cloud, and obtain datasets that explicitly mention 'university' in their metadata. These results can be further reduced with filters, for example to include only the ones that provide an example resource in the RDF/XML format.

One of the key aspects of the design of the LinkedUp catalogue is that it itself creates a Linked Data resource in addition to the use of Datahub.io. Indeed, once datasets have been identified and registered, basic metadata related to each of them, as well as information about their content, are automatically extracted from Datahub.io and from their SPARQL endpoint. This information is then

---

[7] http://data.linkededucation.org/linkedup/catalog/.

Fig. 2. Screenshots of the Web interface to the LinkedUp Catalogue of Datasets for Education for browsing datasets.

represented in RDF using the VoID vocabulary[8] [1] and made available, in a Linked Data way, through a SPARQL endpoint. It is this SPARQL endpoint that we use and interrogate to analyse the characteristics of existing datasets for education in the next section.

3.2   The State of Linked Data in Education

An initial analysis of an earlier version of the catalogue was shown in [5]. It focused on the connection between datasets through their reuse of common vocabulary elements. The core figures from that paper are reproduced in Fig. 3 below, showing the network of datasets and there partitioning through the common reuse of vocabulary elements, and the most commonly used classes/concepts in these datasets, connected by their co-occurrence.

The current version of the LinkedUp catalogue is however much bigger: It references 56 different datasets, each with their own SPARQL endpoint. Datasets are obtained from a variety of sources. As can be seen from Fig. 4 however, they essentially originate from either universities publishing their own data, or from repositories of educational or research resources. Government open data also contribute significantly to datasets related to education, with for example statistics about the registration and results of educational institutions.

A simple aspect one might want to look at when analysing datasets about education from the LinkedUp Catalogue is the variety of sizes that the datasets represent. Each dataset might in particular be divided into multiple sub-graphs,

---

[8] http://www.w3.org/TR/void/.

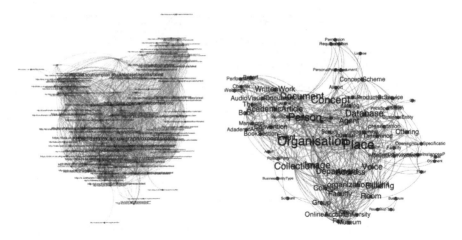

Fig. 3. Dataset (RDF graphs) connected by their reuse of common classes, and common concepts (classes) connected by their co-occurrence (from [5]).

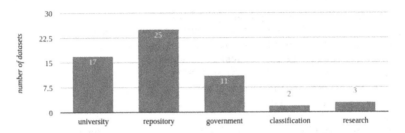

Fig. 4. Number of datasets from different areas in the LinkedUp Data Catalogue.

which might relate to different topics or originate from different sources. As shown in Fig. 5 (which only includes datasets with more than one sub-graph) the number of graphs included in each dataset can vary enormously (from only 1, to several thousands) depending on the way the dataset has been designed and constructed. For example, the biggest one in number of graphs from Fig. 5 (SEEK-AT-WD) is constituted through crowdsourcing, and assigns a different graph to each contribution. Some universities would include in one graph all the information about all the courses they offer, while others might create a graph for the representation of each course. As can be seen however, besides datasets with very large numbers of graphs, or small datasets focusing on a small number of topics, most datasets are structured using 10 to 100 graphs corresponding to different aspects of the data (e.g. modules, resources, people, facilities, etc.).

For information, the chart in Fig. 5 is generated from the results of the following query on the SPARQL endpoint of the LinkedUp catalogue[9]:

---

[9] http://data.linkededucation.org/linkedup/catalog/sparql/.

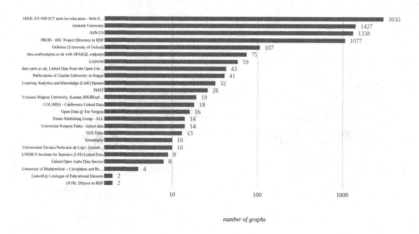

Fig. 5. Number of RDF graphs for datasets with more than one graph (log scale).

```
select distinct ?t (count(?sg) as ?n) where {
  graph <http://data.linkededucation.org/linkedup/catalogue/void> {
    ?d a <http://rdfs.org/ns/void#Dataset>.
    ?d <http://rdfs.org/ns/void#sparqlEndpoint> ?x.
    ?d <http://purl.org/dc/terms/title> ?t.
    ?d <http://rdfs.org/ns/void#subset> ?sg
}} group by ?t order by desc(?n)
```

Similarly we can look at the size of dataset through comparing the number of classes and properties they use. To an extent, the number of classes gives an idea of the variety of the dataset, while the number of properties indicates a notion of richness. Figure 6 shows the number of classes and properties of each dataset that refer to at least 1 class in any of their graphs. Once again, it is clear that there is a wide variety across the datasets of the LinkedUp Catalogue. Several datasets cover information about a very small number of classes (sometimes only one), meaning that the focus on a specific and restricted type of data objects (for example educational resources). Amongst these focused datasets, some still use a comparatively large number of properties, indicating that the information available about each data object in those datasets can be expected to be rich. In the other end of the spectrum are datasets with a very large number of classes, which can include dataset representing a thesaurus or classification, where each topic is a class, or others that generate/use very granular classes to represent the different types of objects they represent.

To generate the data at the basis of Fig. 6, we used the following SPARQL query to the SPARQL endpoint of the LinkedUp Data Catalogue:

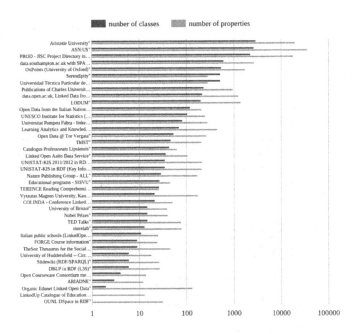

number of classes and properties

Fig. 6. Number of classes and number of properties for datasets of the LinkedUp Data Catalogue (log scale).

```
select distinct ?t (count(distinct ?cp) as ?nc)
                    (count(distinct ?pp) as ?np) where {
  graph <http://data.linkededucation.org/linkedup/catalogue/void> {
    ?d a <http://rdfs.org/ns/void#Dataset>.
    ?d <http://rdfs.org/ns/void#sparqlEndpoint> ?x.
    ?d <http://purl.org/dc/terms/title> ?t.
    {{?d <http://rdfs.org/ns/void#subset> ?sg.
      ?sg <http://rdfs.org/ns/void#classPartition> ?cp.
    ?sg <http://rdfs.org/ns/void#propertyPartition> ?pp}
    UNION
    {?d <http://rdfs.org/ns/void#classPartition> ?cp.
    ?d <http://rdfs.org/ns/void#propertyPartition> ?pp }}
}} group by ?t order by desc(?nc)
```

To really understand the way Linked Data is used to represent data for education, a possible way is to consider an overview of the kind of content they consider. This can especially be done through looking at the types of the data objects that they include, i.e. the classes that they employ to model the data. Looking at Fig. 7, it is interesting to see that, amongst the most popular classes in the datasets, the first is the one used to model people in the FOAF vocabulary[10].

---

[10] http://xmlns.com/foaf/spec/.

Indeed, it appears that many of the educational datasets put a strong emphasis of the way people are involved in education, considering in particular university staff and the way they relate to the educational institutions and organisations they are working with (represented, a bit below in the list, by the Organization and Institution in the FOAF and AIISO[11] vocabularies respectively). Unsurprisingly too, several of the most popular classes relate to the formats in which the data is modeled, including RDF, OWL and DataCube. Again unsurprisingly considering the many datasets originating from repositories (as shown in Fig. 4), most of the remaining classes in Fig. 7 relate to different forms of educational resources or resources that can be used for education, including Document (from FOAF), Article and Book (from the BIBO ontology[12]).

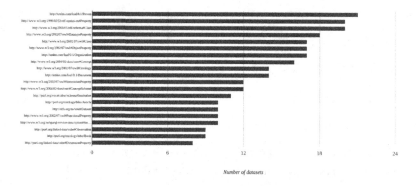

Fig. 7. 20 most common classes amongst the datasets in the LinkedUp Data Catalogue (in number of datasets).

Similarly to classes, the most common properties used in the datasets is an indication of the focus of the content of datasets included in the LinkedUp catalogue. They however give a slightly different picture, as they do not indicate what kind of objects are represented in the data, but what are the dimensions, attributes or indicators most commonly used to describe them. As can be seen in Fig. 8, besides the properties associated with data formats (RDF, etc.), the majority of the most popular properties relate to the modelling of basic metadata attributes of resources, with the Dublin Core vocabulary[13] (for title, description and author for example) as well as to the authors of such resources (for example, the property creator from Dublin Core). Following this, and the conclusion from Fig. 7 that many datasets describe people, we can find amongst the most popular properties also the ones to describe the basic contact information of people, including names, homepages, etc.

The query at the basis of Fig. 7 is described below, and can be straightforwardly adapted to obtain the data at the basis of Fig. 8.

---

[11] http://vocab.org/aiiso/.
[12] http://bibliontology.com/.
[13] http://dublincore.org/.

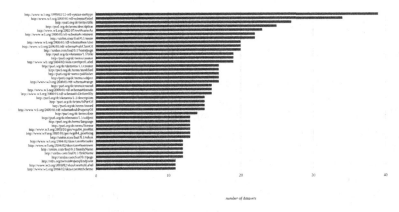

Fig. 8. 40 most common properties amongst the datasets of the LinkedUp Data Catalogue (in number of datasets).

```
select distinct ?c (count(distinct ?d) as ?n) where {
    graph <http://data.linkededucation.org/linkedup/catalogue/void> {
        ?d a <http://rdfs.org/ns/void#Dataset>
        {{?d <http://rdfs.org/ns/void#subset> ?sg.
          ?sg <http://rdfs.org/ns/void#classPartition> ?cp.
          ?cp <http://rdfs.org/ns/void#class> ?c}
        UNION
          {?d <http://rdfs.org/ns/void#classPartition> ?cp.
          ?cp <http://rdfs.org/ns/void#class> ?c} }
    }
} group by ?c order by desc(?n) limit 20
```

4   Use and Future of Linked Data in Education

The analysis described above gives a general overview of the state of Linked Data in education, of the way it is being used and how it can grow further. Indeed, these datasets represent pioneering initiatives that should carry on growing, and through understanding the expending practices of Linked Data in education within these dataset, other can learn from them and have an easier entry point to join the Linked Education CLoud. A key for this to happen however is for these practices to be better shared. Indeed, another element shown in the analysis above is that many of the initiatives at the origin of the considered datasets have been developed in isolation from each other, with different modeling principles, designs and vocabularies being used. While it is the nature of Linked Data (and to an extent of the Web) that it should allow this kind of distribution, some concertation is still required to ensure that the resulting datasets can be used jointly in a way which is sufficiently cohesive (see [5]). Several initiatives have emerged that are trying to address this, among which LinkedUniversities.org, LRMI[14] and the W3C Open Linked Education community group[15].

An important aspect of the state of adoption of Linked Data principles and technologies in education that is not addressed in this chapter is the way it is being used. Indeed, we focus here on the data available and the way it is being modelled, and therefore mostly on the data publication process. The consumption of Web Data for education was actually the main objective of the LinkedUp project, as illustrated by the LinkedUp Challenge: A series of application development competition to encourage the creation of innovative solutions in teaching and learning through the use of Web Data[16]. The result is several dozens of applications at various stages of maturity. These, as well as other examples, show how some areas are emerging as the key applications of Linked Data in education, from the basic management and sharing of data in educational institutions (see for example [4]) to recommendation (see for example [6]).

On of such area which is generating increasing interest is Learning Analytics. Learning Analytics is about the processing of data about learners and their environments for the purpose of understanding and optimising learning (see for example Ferguson, 2012). A lot of both the research-oriented and the practical work in this area is dedicated to the methods employed for collecting, analysing, mining or visualising such data in relation to various levels of models of learning, from the basic information models used to structure the data, to the cognitive models that are expected to be reflected in the learners activity patterns found in the data. It therefore about the way to make sense of raw data in terms of the learners experience, behaviour and knowledge, and Linked Data could represent an approach for the collection, integration and dissemination of such data (see dAquin et al., 2014).

References

1. Alexander, K., Hausenblas, M.: Describing linked datasets - on the design and usage of void, the vocabulary of interlinked datasets. In: Linked Data on the Web Workshop (LDOW 09), in Conjunction with 18th International World Wide Web Conference (WWW 2009) (2009)
2. Brickley, D., Guha, R.V.: RDF vocabulary description language 1.0: RDF schema. W3C recommendation (2004)
3. Daga, E., d'Aquin, M., Adamou, A., Brown, S.: The open university linked data - data.open.ac.uk. Semantic Web Journal - Interoperability, Usability, Applicability (2015)
4. d'Aquin, M.: Putting linked data to use in a large higher-education organisation. In: Interacting with Linked Data at Extended Semantic Web Conference, ESWC (2012)
5. d'Aquin, M., Adamou, A., Dietze, S.: Assessing the educational linked data landscape. In: ACM Web Science (2013)

---

[14] http://www.lrmi.net.

[15] https://www.w3.org/community/opened/.

[16] http://linkedup-challenge.org.

6. d'Aquin, M., Allocca, C., Collins, T.: Discou: A flexible discovery engine for open educational resources using semantic indexing and relationship summaries. In: Demo at International Semantic Web Conference, ISWC (2012)

7. d'Aquin, M., Dietze, S.: Open education: A growing, high impact area for linked open data. ERCIM News, (96) (2014)

8. Guy, M., M., d'Aquin, S. Dietze, H. Drachsler, E. Herder, E. Parodi.: Linkedup: Linking open data for education. Ariadne, (72) (2014)

9. Heath, T., Bizer, C.: Linked Data: Evolving the Web Into a Global Data Space. Synthesis Lectures on the Semantic Web: Theory and Technology, 1st edn. Morgan and Claypool, San Francisco (2011)

10. Klyne, G., Carroll, J.J.: Resource description framework (RDF): Concepts and abstract syntax. W3C recommendation (2006)

11. McGuinness, D.L., Van Harmelen, F.: OWL web ontology language overview. W3C recommendation (2004)

12. PrudHommeaux, E., Seaborne, A.: SPARQL query language for RDF. W3C recommendation (2008)

Educational Linked Data on the Web - Exploring and Analysing the Scope and Coverage

Davide Taibi[1(✉)], Giovanni Fulantelli[1], Stefan Dietze[2], and Besnik Fetahu[2]

[1] Istituto per le Tecnologie Didattiche, Consiglio Nazionale delle Ricerche, Palermo, Italy
{davide.taibi,giovanni.fulantelli}@itd.cnr.it
[2] L3S Research Center, Hannover, Germany
{dietze,fetahu}@l3s.de

Abstract. Throughout the last few years, the scale and diversity of datasets published according to Linked Data (LD) principles has increased and also led to the emergence of a wide range of data of educational relevance. However, sufficient insights into the state, coverage and scope of available educational Linked Data seem still missing. In this work, we analyse the scope and coverage of educational linked data on the Web, identifying the most significant resource types and topics and apparent gaps. As part of our findings, results indicate a prevalent bias towards data in areas such as the life sciences as well as computing-related topics. In addition, we investigate the strong correlation of resource types and topics, where specific types have a tendency to be associated with particular types of categories, i.e. topics. Given this correlation, we argue that a dataset is best understood when considering its topics, in the context of its specific resource types. Based on this finding, we also present a Web data exploration tool, which builds on these findings and allows users to navigate through educational linked datasets by considering specific type and topic combinations.

Keywords: Dataset profile · Linked data for education · Linked data explorer

1 Introduction

The diversity of datasets published according to Linked Data (LD) [5–7] principles has increased in the last few years and also led to the emergence of a wide range of data of educational relevance [18]. These include open educational resources metadata, statistical data about the educational sector, video lecture metadata or university data about courses, research or experts [2]. Initial efforts to collect and catalogue such datasets have been made through initiatives such as the LinkedUp Data Catalog[1] or related community initiatives[2].

---

[1] http://data.linkededucation.org/linkedup/catalog/.
[2] These include, for instance, http://linkededucation.org, http://linkeduniversities.org or the recently established W3C Community Group on Open Linked Education (http://www.w3.org/community/opened/).

D. Mouromtsev and M. d'Aquin (Eds.): Open Data for Education, LNCS 9500, pp. 16–37, 2016.
DOI: 10.1007/978-3-319-30493-9_2

However, the state, coverage and scope of available educational Linked Data have not been widely investigated. Here, in particular questions about the represented resource types, such as, resource metadata or information about organisations or people, and topics are of crucial relevance to shape a better understanding about the state of educationally relevant Linked Data on the Web. Also identifying a dataset containing resources related to a specific topic is, at present, a challenging activity. Moreover, the lack of up-to-date and precise descriptive information has exacerbated this challenge. The mere keywords-based classification derived from the description of the dataset owner is not sufficient, and for this reason, it is necessary to find new methods that exploit the characteristics of the resources within the datasets to provide useful hints about topics covered by datasets and their subsequent classification.

In this direction, authors in [1, 3] proposed an approach to create structured metadata to describe a dataset by means of topics, defined as DBpedia categories, where a weighted graph of topics constitutes a dataset profile. Profiles are created by means of a processing pipeline[3] that combines techniques for datasets resource sampling, topic extraction and topic ranking. Topics have been extracted by using named entity recognition (NER) techniques, where topics are ranked, respectively weighted, according to their relevance using graph-based algorithms such as PageRank, K-Step Markov, and HITS.

The limitations of such an approach are related mainly to the following aspects. First, the meaning of individual topics assigned to a dataset can be highly dependent on the type of resources they are attached to. Also, the entire topic profile of a dataset is hard to interpret if categories from different types are considered at the same time. As an example of the first issue, the same category (e.g. "Technology") might be associated to resources of very different types such as "video" (e.g. in the Yovisto Datset[4]) or "research institution"(e.g. in the CNR dataset[5]). Concerning the second issue, the single topic profile attached for instance to bibliographic datasets (such as: the LAK dataset[6] or Semantic Web Dog Food[7]) - in which people ("authors"), organisations ("affiliations") and documents ("papers") are represented– is characterized by the diversity of its categories (e.g. DBpedia categories: Scientific_disciplines, Data_management Information_science but also Universities_by_country, Universities_and_ colleges). Indeed, classification of datasets in the LD Cloud is highly specific to the resource types one is looking at. While one might be interested in the classification of "persons" listed in one dataset (for instance, to learn more about the origin countries of authors in DBLP), another one might be interested in the classification of topics covered by the documents (for instance disciplines of scientific publications) in the very same dataset.

---

[3] http://data-observatory.org/lod-profiles/profiling.htm.

[4] http://www.yovisto.com/.

[5] http://data.cnr.it/.

[6] http://lak.linkededucation.org.

[7] http://data.semanticweb.org.

In this paper, we aim at providing a systematic assessment of educational Linked Data which consider both, represented topics as well as resource types and their correlations. Questions of interest are:

1. Which types and topics are covered by existing educational Linked Data?
2. What are the central topics covered for particular types (e.g. Open Educational Resources metadata)?
3. Are certain topics underrepresented for certain types, or vice versa?

The approach we propose overcomes the limitations described above by considering the topic profiles defined in [3] in the context of the resource types they are associated with. However, the schemas adopted by the datasets of the LD cloud are heterogeneous, thus making it difficult to compare the topic profiles across datasets. While there are many overlapping type definitions representing the same or similar real world entities, such as "documents", "people", "organization", type-specific profiling relies on type mappings to improve the comparability and interpretation of types and consequently, profiles. For this aim the explicit mappings and relations declared within specific schemas (for instance, *foaf:Person* being a subclass of *foaf:Agent*) as well as across schemas (for instance through *owl:equivalentClass* or *rdfs:subClassOf* properties) are crucial.

While relying on explicit type mappings, we have based our work on a set of datasets where explicit schema mappings are available from earlier work [2]. This includes education-related datasets identified by the LinkedUp Catalog[8] in combination with the dataset profiles generated by the Linked Data Observatory[9]. While the latter provides topic profiles for the majority of LOD datasets, the LinkedUp Catalog contains explicit schema mappings which were manually created for the most frequent types in the LinkedUp Catalog.

The next Section provides a broad overview on the educational Linked Data from a perspective that highlights the relations with the Open Educational Resource world; then, we provide a thorough state of the art assessment of the coverage and scope of educational Linked Data in Sect. 3, which answer aforementioned questions. In addition, we introduce an interactive explorer of educational Linked Data, in Sect. 4, which aims at providing a resource type-specific view on categories associated with the datasets in the LinkedUp Catalog.

2   Resources for Education: Linked Data and OER

The Semantic Web, and specifically the possibility to publish data on the Web and connect them through links (i.e. the Linked Data model), represents one of the most significant evolution of the Internet, after the idea of the Web itself.

---

[8] http://data.linkededucation.org/linkedup/catalog/.

[9] http://data-observatory.org/lod-profiles.

From an educational point of view, both the human-readable and navigable structure of the Web pages and the machine-processable datasets of LD have opened up incredible potentials for the implementation of new and effective pedagogical paradigms [4].

The hypertextual organization of information and knowledge of the Web has influenced not only the ICT-based educational projects worldwide, but also the publication of traditional school textbooks, where anchor-like notes appear throughout a book for immediate references to other related concepts.

In general, the more evident opportunity of the Web for education is a very basic one, yet extremely important for education: the possibility to publish information that everybody can access and use to develop knowledge.

Some years after the birth of the Web, under the pressure of economical, philanthropic and pedagogical emergencies, the idea to exploit materials published on the Web for educational purposes brought to the development of the Open Educational Resources (OER) movement. Since then, hundreds of OER repositories have populated the Web with resources designed for education.

From a pedagogical perspective, OER have solved some critics related to the use of Web pages for education, such as the lack of a pedagogical structure to present information, or the difficulties in identifying the pedagogical scope of a resource published on the Web. However, the OER movement does not exclude more general resources accessible through the Web, provided that they are included into usage patterns designed according to pedagogical criteria. Furthermore, the spectrum of OER is really wide, ranging from resources produced by academic or educational committed institutions to user- or crowd-generated resources. This variety of OER is reflected in the many definitions of OER that can be found in the literature [13–17].

In spite of their tremendous influence on education, OER have also shown some limits; amongst the others:

- The lack of a sole standard for OER and repositories, which has fragmented the offer of OER on the Web [8];
- The complexity in handling direct links between OER and, consequently, in finding semantically related resources.
- The impossibility to guarantee metadata interoperability, due to the proliferation of educational metadata schemas [9];
- The impossibility to deal with the vast availability of education-related data on the web.

The Linked Open Data model offers new solutions for educational resources, partly solving some of the OER limits, still representing a paradigm that complements the OER one, and does not substitute it.

Amongst the OER issues that can be solved by the LOD approach:

- LOD are interlinked by definition; consequently, algorithms can automatically identify semantically related resources; in the case of OER, it was necessary to develop a semantic layer to describe OERs;

– Federated query can be used in order to find resources belonging to distinct datasets; as far as OER are concerned, this was only possible if OER repositories were federated, e.g. through the OAI PMH which allows the exposition of metadata through a common protocol;
– LOD provide the solutions to publish education-related data on the web.

For these reasons, the interest of the educational community in LOD has developed over the years, even sustained by the growing availability of resources published in the Linked Data format, which has raised from 12 in 2007 up to 570 in 2014[10].

The first applications of Linked Data to education focused on the potentials of LD to solve interoperability issues in the field of TEL (Technology Enhanced Learning). In the mEducator project [12], data from a number of open TEL data repositories has been integrated, exposed and enriched by following LD principles. Afterwards, more and more attention has been paid to the increasing availability of datasets on the Web, and particularly to the presence of educational information in the linked data landscape.

The LinkedUp project has explicitly aimed at the educational exploitation of Linked Open Data, and has distinguished two types of linked datasets: datasets directly related to educational material and institutions, including information from open educational repositories and data produced by universities; datasets that can be used in teaching and learning scenarios, while not being directly published for this purpose.

Therefore, the approach followed by the LinkedUp project enhances the general principle of the OER movement that not only resources explicitly developed for educational purposes can be used in educational patterns.

From one hand, this is an essential advantage of open education in general; however, it amplifies some drawbacks that could hinder the potentials of LOD in education:

– *Which datasets and resources can be employed in educational contexts?* A similar challenge has been already addressed in the OER world. However, this task presents a higher degree of complexity for LOD, since the OER movement focuses on the development of content on pedagogical principles, while generally there is no pedagogical theory behind the publication of a dataset, and classifying them becomes more complicated.
– *How datasets (and their resources) should be described in order to facilitate their search (and pedagogical exploitation)?* This issue shows one of the main difference between OERs and LOD. While a bad-described OER can be easily visited by the end user in order to check if it is suitable for a specific educational project, a bad-described dataset can be hardly analysed by the end user, and the risk that the dataset will be ignored is very high.

For these reasons, specific classification mechanisms as the ones described in this chapter, which highlight the key elements of a dataset, together with search tools based on the, are extremely important to fully exploit the potentials of LOD in education.

---

[10] source: http://lod-cloud.net/.

3   Analysing the Coverage of Educational Linked Data

In this section, we present the actual analysis of educational Linked Datasets on the Web, taking into account both topics as well as resource types.

3.1   Data and Method

Topic annotations are provided in the form of DBpedia categories for the majority of LD datasets, available from the topic profiles[11] dataset, further described in [3]. A topic profile connects a dataset with the topics extracted from the analysis of resource samples. Since topics are ranked, a topic profile can be seen as a weighted dataset-topic graph. As such a, topic profile provides a comprehensive overview of the topic coverage of individual datasets. Analysed across a specific set of datasets - as carried out in this work - topic profiles provide insights into the coverage of such a set of datasets.

While topic annotations are obtained from analysing resources of a particular type, the semantics of the topic can best be interpreted when considering the type of the resource. As an example, if the topic "Biology" is associated to a resource of type *foaf: Document*, for instance, a scholarly publication, it indicates that this particular resource is related to biological aspects. In case the "Biology" topic is associated to a *foaf: Organization* resource, it is likely referred to a Biology department of a university. Next to such differences in interpreting topics, the nature of DBpedia categories also differs significantly across different types. For instance, while actual document-related types usually are related to topics which indicate some form of subject or domain (such as "Biology"), resources which represent some notion of organisation or person usually are characterised through some broader categorisations, such as "Academic_institutions" or "People_from_Athens". These fundamental differences are important to understand the nature of dataset topic profiles and to motivate our adopted methodology.

Since our work considers the investigation of both, topics and types, we use as additional data source the LinkedUp Catalog[8]. Our research investigates 21 datasets, which is precisely the set of datasets existing in both collections the LinkedUp Catalog and the Dataset Topic Profiles, as only for these both topic profile and resource type mapping annotations were available. The complete list of selected datasets is shown in Table 1. As explained by Fetahu et al. in [3], topic profiles are generated based on resource samples, where the applied sampling strategies did take into account factors such as the population size of respective types leading to different sample sizes across different datasets. Table 1 indicates both the total amount of data and the characteristics of the automatically computed sample.

The analysis of the relationships between datasets, topics and resource types - aimed at providing a response to the research questions posed above - has been undertaken exploiting network analysis theories and methods. Indeed, the connections between the three investigated notions can be represented by networks, in which

---

[11] http://data.l3s.de/dataset/linked-dataset-profiles.

Table 1. Datasets, resources and resource types

Dataset	Total data		Sampled data		
	#Types	#resource	# Types	#resource	# Categ.
asn-us	29	7494200	3	10000	2128
Colinda	21	17006	9	1985	479
data-cnr-it	120	485977	7	29768	2702
data-open-ac-uk	134	386291	7	36668	1979
education-data-gov-uk	99	315632	42	18712	2510
educationalprograms_sisvu	27	104238	22	12627	2144
gesis-thesoz	9	48532	4	1176	487
hud-library-usagedata	6	904747	1	10000	2300
l3s-dblp	6	15514	3	9368	943
lak-dataset	14	13688	3	10000	1496
linked-open-aalto-data-service	22	373553	12	17598	1543
Morelab	13	244	9	890	206
open-courseware-consortium-metadata-in-rdf	4	22850	4	29369	2723
organic-edunet	1	11093	1	847	559
Oxpoints	142	73655	30	8649	1529
publications-of-charles-university-in-prague	258	14324	15	658	197
seek-at-wd-ict-tools-for-education-web-share	556	13502	37	9938	1624
unistat-kis-in-rdf-key-information-set-uk-universities	35	371737	9	39684	556
universitat-pompeu-fabra-linked-data	39	5778	13	1617	312
university-of-bristol	15	240179	12	22572	2450
Yovisto	8	549986	8	5605	2122

the elements are nodes and their relationship are edges. Specifically the analysis of the relationships has been conducted by considering:

- the network representing the relationships between datasets mediated by categories/topics
- the network representing the relationships between datasets mediated by resource types
- the network representing the relationships between resource types mediated by categories/topics

These networks have been represented by using the Open Source software Gephi[12]. Due to the high number of categories connected to certain datasets (as shown in Table 1), dataset profiles have been filtered by selecting for each dataset the top 100 categories with the highest relevance score. Exploiting the insights gained from such networks, we can identify the particular type/topic coverage of educational LD datasets, corresponding gaps, and the correlation of educational resource types and topics.

3.2 Analysing Topic Coverage - the Dataset-Category-Graph

Representing datasets and categories, i.e. topics, as a weighted graph allows us to analyse the topic coverage of assessed datasets and their proximity topic-wise. In particular,

---

[12] http://gephi.github.io/.

a dataset is connected with the corresponding category depending on its topic profile. Indirect relationships among datasets emerge through shared or connected categories.

The nodes of this network represent datasets and categories. In particular, a dataset is connected with the correspondent category depending on its topic profile. For this reason there is not a direct connection between two datasets, but the categories act as indirect links by connecting datasets that share the same categories. Representing datasets and categories, i.e. topics, as a weighted graph allows us to analyse the topic coverage of assessed datasets and their proximity topic-wise. In particular, a dataset is connected with the corresponding category depending on its topic profile. Indirect relationships among datasets emerge through shared or connected categories.

The color and the dimension of the nodes are related to the metrics of the network which were calculated. In particular, the color gradient is related to the degree of a node (a darker node has a higher degree) while the dimension is proportional to the *betweenness centrality* measure. A detailed view of the graph is shown in Fig. 2. Next to other measures that indicate the importance of the nodes based on their topology, the betweenness centrality of a node is calculated by considering the number of the shortest paths from all pairs of nodes that pass through the node.

Table 2 reports the list of the top ten most connected categories in the datasets under investigation by taking into consideration the number of datasets. Note that while each topic is a DBpedia category, we omitted the DBpedia namespace (http://dbpedia. org/category/) from the listing. The number of datasets sharing the specific category is also reported.

Table 2. Top 10 categories according to their number of occurrences in distinct dataset profiles

Category	# dataset
Academic_disciplines	19
Applied_sciences	19
Applied_disciplines	18
Applied_mathematics	18
Artificial_intelligence	18
Areas_of_computer_science	16
Formal_sciences	16
Interdisciplinary_fields	16
Computing	16
Biology	16

Ranking categories according to the number of resources they are associated with, a different set of top-10 categories emerges (Table 3). The number of datasets sharing the specific category is also reported.

The categories reported in Table 3 highlight the heterogeneity of the dataset resources: categories representing actual disciplines (such as *Biology*, *Computing*, as extracted from Open Educational Resources or video lectures) as well as categories related to institutions (such as *Academic_institutions*) are represented in the list. This overview already demonstrates the strong impact of the resource type (eg *foaf:*

Table 3. Most represented categories in the sampled resources

Category	# dataset	# resources	#types
Applied_sciences	19	3581	81
Computing	16	2778	92
Academic_disciplines	19	2328	68
Biology	16	2068	56
Digital_technology	12	2012	51
Education	14	1855	66
Academia	15	1668	63
Academic_institutions	14	1625	54
Interdisciplinary_fields	16	1574	57
Society	12	1476	60

Document or *foaf:Organisation*) on the associated categories, an observation which motivated parts of the following investigations and an explorative browser described in [11] and Sect. 4.

In the network of Fig. 1 the resource type is not considered, thus two datasets can be connected even if they are collecting different types of resources such as information about institutions, learning materials or scientific publications. In this way, the description of each dataset is not only based on keywords and description provided by dataset authors, but useful hints are also provided about the topics to which the resources of a dataset are connected with. Moreover, the network allows to identify clusters of dataset containing resources related to similar topics.

In order to investigate furthermore the influence of the dataset heterogeneity, the following four datasets containing resources of different nature has been selected:

- The LAK Dataset[13] providing scholarly papers in the Educational Data Mining and Learning Analytics research fields [10].
- The data.cnr.it[14] providing information about the National Research Council of Italy institutes.
- The course descriptions contained in the Linked Data endpoint of The Open University UK[15].
- The L3S DBLP[16] dataset that collects papers related to computer science discipline.

All these datasets provide resources for different educational purposes. In fact, the CNR dataset describes the organization level of this institution, providing information about buildings and persons; the LAK dataset contains the description and content of scientific publications, information related the authors of the papers and the organization they are affiliated with; similar data types are contained in the L3S DBLP dataset though covering a broader research field.

---

[13] http://data.linkededucation.org/request/lak-conference/sparql.

[14] http://data.cnr.it/sparql-proxy/.

[15] http://data.open.ac.uk/query.

[16] http://dblp.l3s.de/d2r/sparql.

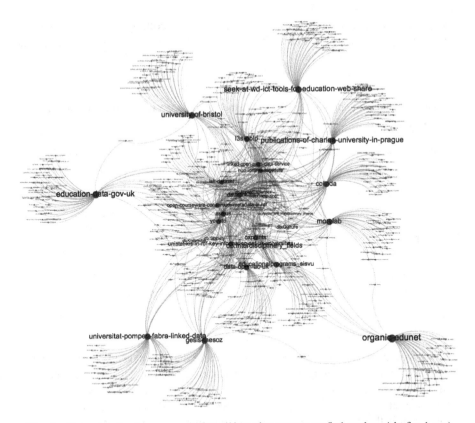

Fig. 1. Dataset and category graph (http://data-observatory.org/led-explorer/ch_fig_1.svg)

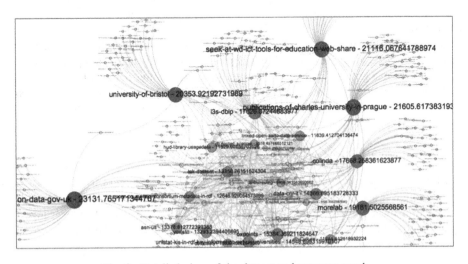

Fig. 2. Detailed view of the dataset and category graph

Finally, the part of The Open University (UK) dataset collects learning materials in different disciplines. Table 4 reports for these four datasets the associated categories with the highest relevance scores.

Table 4. Comparing top-k categories for four datasets of the Linked Educational cloud

k	Lak Dataset	Data.Cnr.it	data-open-ac-uk	L3S DBLP
1	Applied_sciences	Computing	Scientific_disciplines	Data_management
2	Educational_organizations	Interdisciplinary_fields	Interdisciplinary_fields	Information_retrieval
3	Academic_institutions	Data	Applied_sciences	Education
4	Educational_assessment_ and_evaluation	Information_technology	Science	Digital_media
5	Accreditation	Scientific_disciplines	Academic_disciplines	Book_websites
6	Education	Society	Knowledge	Archives
7	Applied_disciplines	Social_sciences	Natural_sciences	Digital_libraries
8	Computing	Data_management	Chemistry	Electronic_publishing
9	Academic_institutions	Academic_disciplines	Physical_sciences	Digital_library_projects
10	Digital_technology	Project_management	Education	Educational_projects
11	Academic_disciplines	Biology	Psychology	Computing_and_society
12	Computer_science	Land_Management	Branches_of_philosophy	Online_content_distribution
13	Data	Computer_data	Biology	Bibliographic_databases

It is important to highlight that even though these datasets have in common several categories, they contain different types of resources. Therefore, clustering mechanism based only on the categories shared by the dataset are not precise. For example, in the selected datasets, the *Biology* category are related with the data.cnr.it and the Open UK datasets even if these two datasets contain resources of entirely different types.

In Fig. 3 the effect of considering the resource type in the analysis of the relationships between dataset and categories is shown, with respect to the four datasets. Indeed, when the *foaf:Document* is considered only the data.cnr.it and lak-dataset are connected.

By considering the *foaf:Agent* the lak-dataset and l3 s_dblp shared categories, finally by considering the *aiiso:KnowledgeGrouping* only the data-open-ac-uk is connected to categories. This result is clearly related to the nature of the datasets analysed. The influence of resource type on dataset relationships is detailed in the next section.

3.3   Resource Type Coverage - the Dataset-Type-Graph

The type of the resources plays a key role to guide the exploration of the datasets and furthermore it is a strong indicator for the connectivity between datasets. Therefore, the influence of the resource types in the relationships between datasets has been investigated further. The Table 5 reports the ten resource types most frequently occurring across the 21 datasets under investigation.

Fig. 3. The effect of resource types on topic connections between datasets (a) all resource types; (b) foaf:Document; (c) foaf:Agent; (d) aiiso:KnowledgeGrouping (http://data-observatory.org/led-explorer/ch_fig_3{a,b,c,d}.svg)

The relationships between datasets and resource type have been analysed by means of the graph in Fig. 3 where nodes are resource types and datasets, an edge connects a dataset with a resource type if the dataset contains resources of that particular type.

In Fig. 4, four disconnected networks are represented. This is due to the fact that some datasets use specific vocabularies for their resources. Moreover, in the construction of this network the semantic relationships between resource types have not been taken into consideration. Consequently, types such as *foaf:Agent* and *foaf:Person* that are related by a *rdfs:subclass* relationship have been considered as distinct types. In order to improve the analysis of the relationships between datasets and resource types both existing mappings and new ones have been introduced. As existing mapping we consider the relationships that can be inferred from explicitly declared statements in the vocabulary used in the datasets. Moreover, in the context of the LinkedUp project[17] a set of additional mappings has been introduced which link equivalent or overlapping types through standard OWL and RDF predicates, such as, *owl:equivalentClass* or *rdfs:subTypeOf*. A detailed description of the process that has led to the definition of these mappings is described in [2].

The following Fig. 5 reports two examples of mappings for the class *foaf:Agent* and *foaf:Document*.

Table 5. Most frequent resource types across educational Linked Datasets

Resource Type	#Datasets
http://xmlns.com/foaf/0.1/Document	5
http://xmlns.com/foaf/0.1/Person	5
http://www.w3.org/2004/02/skos/core#Concept	4
http://xmlns.com/foaf/0.1/Agent	3
http://purl.org/vocab/aiiso/schema#Institution	3
http://xmlns.com/foaf/0.1/Organization	3
http://rdfs.org/ns/void#Dataset	3
http://purl.org/vocab/aiiso/schema#Course	3
http://purl.org/vocab/aiiso/schema#Department	2
http://swrc.ontoware.org/ontology#InProceedings	2

The consideration of the mappings between resource types has made possible the aggregation of nodes representing resource types. In particular, equivalent resource types as well as resource types with subclass relationships have been grouped. The following Table 6 shows the most frequent resource type across the datasets after inference on mappings. In bold, we show the super-type, while the non-bold types indicate the most specific type association.

In Table 6, the resource types highlighted in bold represent the resource type together with all the resource type connected to it by considering the mapping. This table provides a clearer overview of the resources included in the datasets under investigation. Specifically, the most represented resource types (including also all of its subtypes, mapped types, inferred types) are related to *foaf:Document* (for instance, scientific and academic publications, educational resources), *foaf:Agent* (some of the datasets under investigation contain information about organizations, institutions and people) and *aiiso:KnowledgeGrouping*[18], since this class represents resources related to courses, learning modules, and so on. Type mappings across all involved datasets link "documents" of all sorts to the common *foaf:Document* class, "persons" and "organisations" to the common *foaf:Agent* class, and courses and modules to the *aiiso: KnowledgeGrouping* class.

Figure 6 represents the graph of dataset and resource types and their inferred types.

---

[17] http://linkedup-project.eu.

[18] http://purl.org/vocab/aiiso/schema#KnowledgeGrouping.

Fig. 4. Dataset and resource type graph without considering mapping (http://data-observatory. org/led-explorer/ch_fig_4.svg)

The density of this graph is lower than the one not including mappings, but the connectivity is higher. Subgraphs are stronger connected as in the previous network in which mappings are not considered. Indeed, in the network of Fig. 4, the graph density measured by gephi is lower (0.005) than in this case (0,011).

In Table 7, the datasets containing types linked to either or more of the three super-types are listed.

3.4 Type-Topic Correlation

As shown above, the resource type has a strong impact on the nature and semantics of the associated categories. While actual knowledge resources, such as OER, tend to be linked to explicit domains or disciplines, such as *Biology* or *Computer Science*, the range of categories for persons and organisations is of entirely different nature. While topics/categories are always linked to particular resources and their types, the joint analysis of both types and topics is of crucial importance to enable a better understanding of educational Linked Data. Considering the resource types associated

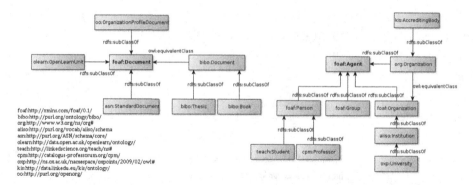

Fig. 5. Mappings between resource types related to foaf:Agent and foaf:Document

Table 6. Most frequent resource types according to their representation in the datasets (mapping considered)

Resource Type	# Datasets
foaf:Agent	**14**
foaf:Person	5
foaf:Organization	3
aiiso:Institution	3
foaf:Agent	3
aiiso:Department	2
foaf:Document	**12**
foaf:Document	5
bibo:Article	2
bibo:Book	2
bibo:Document	2
swrc:Document	2
swrc:InProceedings	2
aiiso:KnowledgeGrouping	**7**
aiiso:Course	3
aiiso:Module	2
courseware:Course	2
skos:Concept	**6**
skos:Concept	4
geo:SpatialThing	**4**
c4 dm:Event	**3**
void:Dataset	**3**

with each topic in the dataset topic profile graph, it has been possible to create a network in which the resource types have been connected with the categories they are related with.

In the graph of Fig. 7, seven groups of nodes are clearly identified. The following table reports the most representative categories related with the most connected

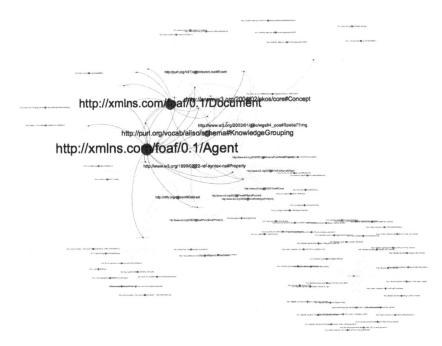

Fig. 6. Datasets and resource type graph considering inferred types (http://data-observatory.org/led-explorer/ch_fig_6.svg)

Table 7. Datasets for the most frequent resource types (considering type mappings)

foaf:Agent	foaf:Document	aiiso:KnowledgeGrouping
morelab	morelab	universitat-pompeu-fabra-linked-data
colinda	organic-edunet	asn-us
publications-of-charles-university-in-prague	asn-us	unistat-kis-in-rdf-key-information-set-uk-universities
lak-dataset	open-courseware-consortium-metadata-in-rdf	oxpoints
university-of-bristol	universitat-pompeu-fabra-linked-data	linked-open-aalto-data-service
13 s-dblp	university-of-bristol	data-open-ac-uk
oxpoints	lak-dataset	educationalprograms_sisvu
education-data-gov-uk	yovisto	
educationalprograms_sisvu	13 s-dblp	
linked-open-aalto-data-service	oxpoints	
unistat-kis-in-rdf-key-information-set-uk-universities	linked-open-aalto-data-service	
	hud-library-usagedata	

resource types in the LinkedUp catalog. In order to enable a better distinction, we particularly consider the most frequent resource types including the resource types associated to them by means of mappings.

Table 8 provides evidence that the resource types related to person and organization (*foaf:Agent*) are more connected to physical places and locations, while resource types related to actual documents (*foaf:Document*) or courses (*aiiso:KnowledgeGrouping*) are representing actual domains and disciplines. For the latter, we observe a strong bias towards topics relating to Computer Science and the Life Sciences. Regarding the *foaf:*

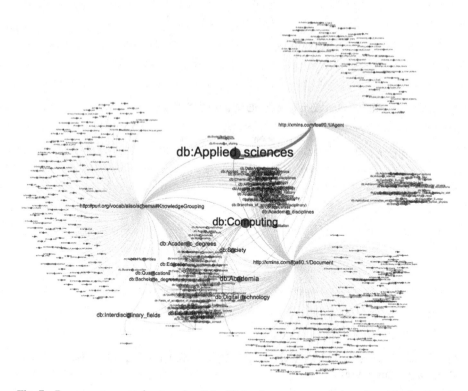

Fig. 7. Resource types and categories (http://data-observatory.org/led-explorer/ch_fig_7.svg)

Organization it is possible to highlight a bias towards the categories related to European area such as: *Cities_in_Europe, Baltic_Sea, Capitals_in_Europe*.

This observation correlates with the general intuition that such topics are also stronger represented in the Linked Open Data cloud in general and might lead to additional research into how to resolve such a topic bias in the future.

This observation can be repeated when analysing a single dataset. For instance in the case of the University of Bristol[19] dataset, that contains resources related to documents and agents, the category distribution among the most frequent resource types/categories is shown in Table 9. The top ten categories according to the score calculated in the profile are reported in the table.

In general, when the categories related to a dataset are analysed, the distribution of the categories is usually heterogeneous if the resource type is not considered. When the resource type is taken into account, so that the types where they originate from (document, person, organization) are identified, the category subgraph pertaining to a specific resource type usually can is more coherent, distinct and connected. This observation can be exploited, for instance, when aiming to automatically map resource types and identifying type relationships, where the associated category overlap or similarity can serve as indicator.

---

[19] http://resrev.ilrt.bris.ac.uk/data-server-workshop/sparql.

Table 8. Most frequent categories for most frequent resource types

foaf:Document		foaf:Agent		aiiso:KnowledgeGrouping	
Applied_sciences	1164	Applied_sciences	1522	Digital_technology	1393
Biology	680	Academic_institutions	533	Computing	1262
Academic_disciplines	656	Academic_disciplines	823	Society	1011
Branches_of_philosophy	624	Educational_organizations	533	Interdisciplinary_fields	793
Chemistry	604	Types_of_organization	523	Education_by_subject	789
Areas_of_computer_ science	593	School_types	520	Academia	717
Education	591	Schools	520	Academic_disciplines	688
Artificial_intelligence	581	Organizations	520	Education	653
Computing	548	Educational_institutions	520	Applied_sciences	648
Branches_of_psychology	548	Educational_buildings	516	Qualifications	591

Table 9. Distribution of top-k categories for the University of Bristol dataset

k	All resource types	foaf:Document	foaf:Agent	
			foaf:Person	foaf:Organization
1	db:Academic_disciplines	db:Academic_disciplines	db:Buildings_and_structures_ by_type	db:Academic_disciplines
2	db:Buildings_and_structures_ by_type	db:Biology	db:Chemistry	db:Buildings_and_structures_ by_type
3	db:Biology	db:Applied_sciences	db:Cities_in_Europe	db:Biology
4	db:Applied_sciences	db:Chemistry	db:Capitals_of_country_ subdivisions	db:Applied_sciences
5	db:Academic_publishing	db:Articles_including_recorded_ pronunciations_(English)	db:Capitals	db:Chemistry
6	db:Academic_publishing_ companies	db:Articles_including_recorded_ pronunciations	db:Arts_in_the_United_Kingdom	db:Business_organizations
7	db:Chemistry	db:Articles_including_recorded_ pronunciations_(UK_English)	db:Capitals_in_Europe	db:Building_engineering
8	db:Cities_in_Europe	db:Academia	db:Cities_in_the_United_States	db:Branches_of_philosophy
9	db:Academia	db:Academic_journals	db:Academic_institutions	db:Anatomy
10	db:Cities_in_South_West_ England	db:Book_publishing_companies_ by_country	db:British_capitals	db:Architecture

4 An Interactive Explorer for Educational Linked Data

In this section, we present the *dataset profile explorer* developed with the aim of supporting an effective re-use of the resources in the educational LD cloud. In particular, the *explorer* makes explicit the topics covered by the datasets even in relation to the types of resources, mainly focused on the domain of educational related datasets. As stated in previous Sect. 3, topic coverage and the type of the resources in this domain assume a key role in supporting the search for content suitable for a specific learning course. The *explorer* allows users to navigate topic profiles associated with datasets with respect to the type of the resource in the dataset.

As shown in Fig. 8, the *explorer*[20] is composed of three panels: the panel at the centre of the screen shows the network of datasets and categories; the panel on the left shows general and detailed descriptions about datasets and categories; the *selection*

---

[20] http://data-observatory.org/led-explorer/.

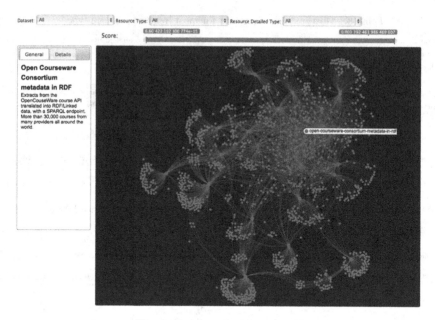

Fig. 8. A screenshot of the demo

panel, placed at the top of the screen, allows users to apply specific filters on the network. In the central panel, green nodes represent datasets while blue nodes represent categories. An edge connects a dataset to a category if the category belongs to the dataset topic profile. In order to draw the network, the *sigmajs*[21] library has been used and the nodes of the network have been displayed using the ForceAtlas2 layout. By clicking on a node (dataset or category), general and detailed descriptions are shown on the left panel. In the case of a dataset, the general description reports the description of the dataset retrieved from the *datahub.io* repository[22]. In the detailed description, the list of the top ten categories (and the related score) associated to the dataset is reported. In the case of a category, the description panel reports the list of datasets which have that category in their profile. The datasets including the category in their top ten list are highlighted in bold.

The selection panel allows users to filter the results according to: dataset, resource type, and resource *sub-type*. The list of datasets is composed by the datasets of the LinkedUp catalog. Regarding the resource type, the *explorer* is focused on three main classes: *foaf:Document*, *foaf:Agent* and *aiiso:KnowledegeGrouping*. As reported in Subsect. 3.4, these three classes are the most represented classes in the datasets, and *foaf:Document* is related to learning material such as: research papers, books, and so on; the *foaf:Agent* resource type has been included to take into account elements such as persons and organizations. The *aiiso:KnowledegeGrouping* is a type representing

---

[21] http://sigmajs.org.

[22] http://datahub.io.

resources related to courses and modules. This initial set of resource types can be easily enlarged by means of configuration settings. The resource sub-type has been included with the aim of refining the results already filtered by resource type. Another filter is related to the score of the relationships between datasets and categories. A slider bar allows users to filter results based on a specific range of the scores, calculated by the linked dataset profiling pipeline [3]. The filters on datasets, resource types and resource sub-types can be combined and, as a result, only the portion of the network consistent with the filter selection is highlighted. Even though the *explorer* has been tested with an initial group of datasets of the LinkedUp Catalog, it can be configured in order to extend the number of datasets covered. Moreover, the *explorer* can cover also datasets on different fields provided that the dataset topic profile is available, thus extending the application of the proposed approach to several fields.

5   Conclusion

In this work, we have provided an analysis of the coverage of educational Linked Data on the Web and an investigation of the inherent correlations between types, topics and datasets. Only the joint consideration of types and topics allows the non-ambiguous interpretation of topic annotations of datasets. Key findings of our study include:

F1. Educational datasets can best be characterised (profiled) by a combined representation of resource types and categories as part of dataset profiles.

F2. The nature of categories differs significantly depending on the resource types they are associated with. In other words, the distinct subgraphs of the DBpedia category graph characterise resources of very distinct types.

F3. Educational and presumable cross-domain resource types can be characterised by their inherent topic distribution.

F4. Educational resources, i.e. instances which represent some form of educational documents, currently are not equally spread across all disciplines. A topic bias exists towards fields in the area of Computer Science and the Life Sciences.

Our analysis uncovers an inherent topic bias of educational resources represented in datasets, usually focused on disciplines related to *Computer Science* and *Life Sciences*, where for instance, social sciences appear to be underrepresented. While this bias emerged for specific resources, i.e. instances by types which can be summarised as some notion of *document*, including dedicated OER, scholarly papers or audiovisual material, a similar bias was not detected for other types such as organisations or persons. In such cases, a deeper analysis, for instance of the origins and characteristics of represented entities, taking into account background knowledge such as geodata, seems better suited to detect some form of demographic or geographic scope or bias. As shown above, the nature of categories associated with resources of different types differs significantly depending on the respective resource type. For instance, while actual document-related types usually are related to topics which indicate some form of subject or domain (such as "*Biology*"), resources representing some notion of agent, such as organisation or person, usually are characterised through some broader cate-gorisations, such as "*Academic_institutions*" or "*People_from_Athens*".

While this suggests that the subgraphs of the DBpedia category graph tied to specific resources fundamentally differs, we argue that category distributions of resource types might provide a useful means for mapping and aligning types (F3). The intuition is that similar categories are likely to be tied to instances of similar resource types. While type mappings in the educational Linked Data landscape as well as the LinkedUp data catalog currently are mostly created manually by experts, as part of future work we are investigating possibilities to exploit this observation as part of automated type and schema alignment methods.

Acknowledgments. This work has been partially supported by the European Union Seventh Framework Programme (FP7/2007-2013) under grant agreement No 317620– LinkedUp project (http://linkedup-project.eu/). The authors would like to acknowledge networking support by the COST Action IC1302 (KEYSTONE).

References

1. Fetahu, B., Dietze, S., Nunes, B.P., Taibi, D., Casanova, M.A.: Generating structured profiles of linked data graphs. In: 12th International Semantic Web Conference (ISWC2013), Sydney, Australia (2013)
2. D'Aquin, M., Adamou, A., Dietze, S.: Assessing the educational linked data landscape. In: ACM Web Science 2013 (WebSci 2013), Paris, France, May 2013
3. Fetahu, B., Dietze, S., Pereira Nunes, B., Antonio Casanova, M., Taibi, D., Nejdl, W.: A scalable approach for efficiently generating structured dataset topic profiles. In: Presutti, V., d'Amato, C., Gandon, F., d'Aquin, M., Staab, S., Tordai, A. (eds.) ESWC 2014. LNCS, vol. 8465, pp. 519–534. Springer, Heidelberg (2014)
4. Dietze, S., Yu, H.Q., Giordano, D., Kaldoudi, E., Dovrolis N., Taibi, D.: Linked education: Interlinking educational resources and the web of data. In: ACM Symposium on Applied Computing (SAC-2012), Special Track on Semantic Web and Applications (2012)
5. Bizer, C., Heath, T., Idehen, K., Berners-Lee, T.: Linked data on the web (LDOW2008). In: Proceedings of the 17th International Conference on World Wide Web (WWW 2008), April 21-25, 2008, Beijing, China (2008)
6. Bizer, C., Heath, T., Bernes-Lee, T.: Linked data - the story so far. Special Issue on Linked data. International Journal on Semantic Web and Information Systems (IJSWIS) (2009)
7. Heath, T., Bizer, C.: Linked data: evolving the web into a global data space (1st edition). In: Synthesis Lectures on the Semantic Web: Theory and Technology, 1:1, 1–136. Morgan & Claypoo (2011)
8. de Santiago, R., Raabe, A.L.A.: Architecture for learning objects sharing among learning institutions-LOP2P. IEEE Trans. Learn. Technol. **3**, 91–95 (2010)
9. Dietze, S., Sanchez-Alonso, S., Ebner, H., Yu, H.Q., Giordano, D., Marenzi, I., Pereira, N. B.: Interlinking educational resources and the web of data: a survey of challenges and approaches. Emerald Program: Electron. Libr. Inf. Syst. **47**(1), 60–91 (2013). doi:10.1108/00330331211296312
10. Taibi, D., Dietze, S.: Fostering analytics on learning analytics research: The LAK dataset. In: CEUR WS Proceedings vol. 974, Proceedings of the LAK Data Challenge, held at LAK2013 – 3rd International Conference on Learning Analytics and Knowledge (Leuven, BE, April 2013) (2013)

11. Taibi, D., Dietze, S., Fetahu, B., Fulantelli, G.: Exploring type-specific topic profiles of datasets: A demo for educational linked data. In: Poster & System Demonstration Proceedings of 13th International Semantic Web Conference (ISWC 2014), Riva Del Garda, Italy, October 2014
12. Mitsopoulou, E., Taibi, D., Giordano, D., Dietze, S., Yu, H.Q., Bamidis, P., Bratsas, C., Woodham, L.: Connecting medical educational resources to the linked data cloud: The mEducator RDF schema, store and API, in linked learning 2011. In: Proceedings of the 1st International Workshop on eLearning Approaches for the Linked Data Age, CEUR-WS, vol. 717 (2011)
13. UNESCO. Forum on the impact of Open Courseware for higher education in developing countries. Final report. Paris: UNESCO (2002)
14. Geser, G.: Open Educational Practices and Resources. OLCOS Roadmap (2012)
15. Hylén, J.: Open educational resources: Opportunities and challenges. In: Proceedings of Open Education 2006: Community, Culture and Content, pp. 49–63 (2006)
16. Atkins, D.E., Brown, J.S., Hammond, A.L.: A review of the Open educational Re- sources (OER) movement: achievement, challenges and new opportunity. Report to the William and Flora Hewlett Foundation (2007)
17. OECD: Giving Knowledge for free: the Emergence of Open Educational Resources. OECD, Paris (2007)
18. Taibi, D., Fulantelli, G., Dietze, S., Fetahu, B.: Evaluating relevance of educational resources of social and semantic web. In: Hernández-Leo, D., Ley, T., Klamma, R., Harrer, A. (eds.) EC-TEL 2013. LNCS, vol. 8095, pp. 637–638. Springer, Heidelberg (2013)

Applications of Open and Linked Data in Education

ECOLE: An Ontology-Based Open Online Course Platform

Vladimir Vasiliev, Fedor Kozlov[(⊠)], Dmitry Mouromtsev,
Sergey Stafeev, and Olga Parkhimovich

ITMO University, St. Petersburg, Russia
vasilev@mail.ifmo.ru, kozlovfedor@gmail.com, d.muromtsev@gmail.com,
stafeevs@yahoo.com, olya.parkhimovich@gmail.com

Abstract. The chapter presents use cases and architecture of an educational platform built upon educational Linked Open Data. We review
ontologies designed for educational data representation and discuss in
detail the Enhanced Course Ontology for Linked Education (ECOLE)
as an example. Its semantic core opens up a variety of new possibilities
for interaction between an e-learning system and related Web-services
and applications. The other feature of the ontology-based platform is a
flexible structuring and linking of open educational resources. The last
part of the chapter discusses these new possibilities and analyzes trends
in linked learning.

Keywords: Semantic web · Linked learning · Terminology extraction ·
Education · Educational ontology population

1 Introduction

Semantic technologies enable a completely new approach to learn on the Internet by means of semantic agents. And there is a number of initiatives regarding creating and publishing open educational data including five stars datasets.
In many cases search engines and knowledge graphs already provide sufficient
support for basic online education. For more complicated educational scenarios
there are resources like BBC Bitesize[1] where the learner can find educational
content organized by means of ontologies. In our work we address the challenges
of understanding how semantics can help to manage educational data and to
make teaching with electronic materials more personalized with respect to the
skills and knowledge background of a particular user.

To answer these questions we developed an experimental ontology-based
open online course platform ECOLE. Initially ECOLE has been created with an
intent to provide a framework for developing e-learning systems in the Semantic
Web era. But step-by-step we realized the role of ECOLE gradually evolved to
become an Enhanced Course Ontology for Linked Education, or in other words a

---

[1] http://www.bbc.co.uk/education.

© Springer International Publishing Switzerland 2016
D. Mouromtsev and M. d'Aquin (Eds.): Open Data for Education, LNCS 9500, pp. 41–66, 2016.
DOI: 10.1007/978-3-319-30493-9_3

semantic layer for educational resources linking and integration. And as a result ECOLE exists as a semantic core of the e-learning system based entirely on OWL ontologies and semantic technologies. It should not be compared with the most advanced e-learning platforms such as edX, Moodle and others because the purpose of ECOLE is to make electronic education more personal and flexible, to make it possible to reuse existing educational resources and to provide more intelligent interactive teaching and analytical functionality for end users. While the most popular e-learning platforms are learning management systems for electronic and distance education our development is focused on the educational knowledge representation and e-learning analytics.

This chapter is organised as follows: Sect. 1 presents a brief survey of related work and defines the problem of interest. Section 2 describes the ontology development for all educational activities. Section 3 explains some technical details of populating the ontology including natural language processing over educational materials. Section 4 describe the ECOLE system architecture and application illustrated with examples of the student UI and the analytical back-end. Finally, Sect. 5 presents the evaluation results.

1.1   Related Work

There are two kinds of applications based on semantic technologies for educational purposes. The first type of projects are based on the principles of Linked Data and aimed at publishing of research and educational data in RDF format for the purposes of search and exchange of information. Probably the biggest ones are the Linked Universities[2] initiative, an alliance of European universities committed to exposing their public data as Linked Data, VIVO[3] project in US that provides a platform for integrating and sharing information about researchers and their activities and the Open University[4], a distant learning and research university with over 240,000 students. The second type of projects is trying to use semantic data models for managing information inside learning platform for making them more flexible, integrated and interactive. One example of semantics usage in the field of education is mEducator. It is a content sharing approach to medical education based on Linked Data principles. Through standardization, it enables sharing and discovery of medical information [1]. Another example is already mentioned project Bitesize from BBC. One more good example of using semantics to make educational materials reusable and flexible is the SlideWiki [2] system, a collaborative OpenCourseWare platform dealing with presentations and performance assessment. It uses CrowdLearn concept as a comprehensive data model that allows a collaborative authoring of highly structured educational materials and improve its quality by means of crowd-sourcing or co-creation. An original approach to integration of semantic technologies into an educational environment is presented in the work of F. Zablith [3]. The author

---

[2] http://linkeduniversities.org/.

[3] http://vivoweb.org/.

[4] http://www.open.ac.uk/.

describes his results on creation of a semantic Linked Data layer for conceptual connection of courses taught in a higher education program. He also presents applications which show how learning materials can float around courses through their interlinked concepts in eLearning environments. The last three examples are presented later in this book in details.

Having in mind the main challenge of our work and results presented below in this chapter it is important to mention a number of ontologies that already exist in the area of e-learning. In our overview we included several examples that could be classified by its purpose into three groups:

- modeling a structure of a course,
- referencing to some educational resources, and
- linking of particular parts of learning processes.

Probably the most popular ontology for representation of courses and modules is The Academic Institution Internal Structure Ontology (AIISO)[5]. AIISO provides classes and properties to describe the internal organizational structure of an academic institution.

For representation of references in semantic formats there are The Bibliographic Ontology(BIBO)[6] and The Ontology for Media Resources(MA-ONT)[7]. BIBO is used to describe bibliographic resources associated with the course such as books or papers. BIBO provides main concepts and properties for describing citations and bibliographic references. MA-ONT describes a core vocabulary of properties and a set of mappings between different metadata formats of media resources published on the Web. MA-ONT is used to store video lectures and additional video materials.

Finally a good example of linking ontology is the TEACH[8] (Teaching Core Vocabulary). TEACH is one of the most relevant and recent ontologies published in the field of education. It is a lightweight vocabulary providing terms to enable teachers relate things in their courses together. TEACH is based on practical requirements set by providing seminar and course descriptions as Linked Data.

The more complete overview of ontologies and vocabularies for education is presented here http://linkeduniversities.org/lu/index.php/vocabularies/.

We tried to reuse ontologies listed above where it was possible. And in the Sect. 2 detailed description of the use of existing ontologies is given. At the same time a personalization of electronic teaching with respect to the skills and knowledge background of a user is still being an open question. The contribution of ECOLE here is in modeling of domain knowledge, user activity and his/her knowledge assessment on top of courses structure and external educational resources. This include the following key aspects:

---

[5] http://purl.org/vocab/aiiso/schema.

[6] http://purl.org/ontology/bibo/.

[7] http://www.w3.org/ns/ma-ont.

[8] http://linkedscience.org/teach/ns/.

- Boost integrity of a course parts by means of shared domain models and inferred indirect links (see Sect. 2.2).
- Automate an evaluation of knowledge assessment materials with a semantic analyses of lecture coverage by tests (see Sects. 2.3 and 4.3).
- Log a user activity in an e-learning system adjusted to domain knowledge models (see Sect. 2.4) that allows,
- Calculate a user knowledge rate by domain (see Sects. 2.5 and 4.3) that gives links to precise educational materials related to a particular domain concept, or term, that user should repeat in case his/her knowledge rate is low.

Obviously all these aspects are beyond the functionality of traditional e-learning systems and semantic technologies here have a great potential. In the next sections of this chapter we explain our ontology-based approach to solve the problem of personalization of open online education as it was set up in this introductory section.

1.2 Motivation

The major task in developing and maintaining an ECOLE system is choosing and interlinking relevant materials, e.g., creating associations between subject terms (or just terms in the context of his chapters) in lectures, practice and tests. This requires domain ontologies and their population with facts from educational content. When data from different external resources is integrated into the course, it can impair the quality of the content. Therefore one of the goals of ECOLE is to provide tools for tutors to check quality of the course using the relations between elements of the course. Another way to improve the quality of the course is to use students activity in the system. In the ECOLE system any kind of sophisticated statistics can be gathered, e.g. statistics about students' correct/incorrect answers allowing to filter out troublesome terms and topics. Teachers can use this statistics to improve the quality of their courses. Students can use personal statistics to fill their knowledge gaps.

2 Ontology Development

All data in the ECOLE system is stored in RDF using a set of developed ontologies. The data model of the ECOLE system contains three basic data layers: the Domain Data Layer, the Educational Data Layer, and the Activity Data Layer. The layers are linked with each other to support interoperability between variety of resources of the system. The data model is shown in Fig. 1.

The Domain Data Layer contains information about subject fields of education and science. This layer is the core of the ECOLE data model. Its data changes rarely and is gathered from external knowledge bases, taxonomies, and datasets such as DBpedia and Mathematics Subject Classification.

The Educational Data Layer contains educational content for teaching. The layer stores the educational programs, courses, tests, and media resources.

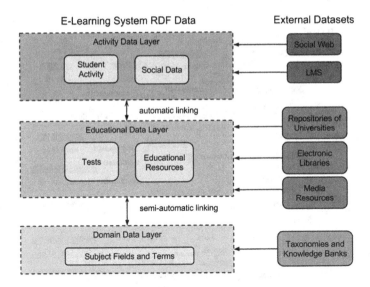

Fig. 1. The data model of ECOLE system

This data is expected to change frequently and can be gathered from repositories of universities, open libraries, and media providers. The entities of the Educational Data Layer can be linked to the entities of the Domain Data Layer manually or automatically using NLP algorithms or automated reasoning.

The Activity Data Layer contains statistical data about system users. The layer stores information about students, their activity in the Learning Management System, and their learning results. The content of this layer changes all the time. It is gathered from the users of Learning Management Systems and various social networks. The entities of Activity Data Layer can be linked to the entities of the Educational Data Layer automatically using the algorithms provided by the Learning Management System.

2.1   The Ontology of Educational Resources

The ontology of educational resources describes relations between courses, modules, lectures, and terms and helps to represent its properties and media content. The original ontology is built on top of uper level ontologies that are commonly used in descriptions of educational resources [5]. These ontologies are shortly described in the Sect. 1.1.

The ontology of educational resources[9] has the following common classes: Course, Module, Lecture, Test, Exam, Practice, Filed, Term, Resource. The ontology contains 32 classes, 42 object properties, and 13 datatype properties. The classes of the developed ontology are shown in Fig. 2.

The most outstanding feature of this ontology is its ability to create direct and indirect interdisciplinary relations between courses [4]. E.g., physics test

---

[9] http://purl.org/ailab/education.

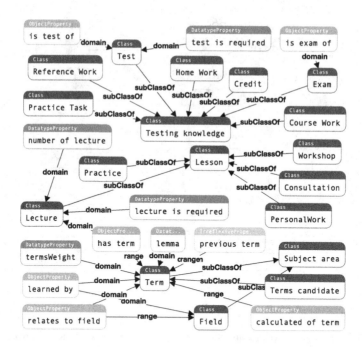

Fig. 2. Main classes of the ontology of educational resources

"Interference and Coherence" includes math terms as well ("vector", "vector product"). Thus, if a student cannot pass this test, the system advises to repeat not only the lecture "Occurrence of Interference" in the Physics course, but also the corresponding lectures from the Vector Algebra course. This is an example of indirect links between physics and vector algebra via the subject terms "vector" and "vector product". An example is shown in Fig. 3.

2.2   Ontology Mapping

We use ontology mapping techniques to support interoperability amongst educational systems based on the developed ontology. Ontology mappings define correspondences between concepts in different ontologies. In this chapter ontology mappings are used to map a concept found in the ontology of educational resources into a view, or a query, over other ontologies. We have chosen the TEACH (Teaching Core Vocabulary) ontology [6] as the target for the ontology mapping purposes and based on its specification we perform ontology mapping manually. Equivalent classes are linked using the `owl:equivalentClass` axiom of the OWL Web Ontology Language [7,8]. Equivalent properties in the two ontologies are linked using the `owl:equivalentProperty` axiom.

The results of ontology mapping between the ontology of educational resources and TEACH ontology are shown in Table 1.

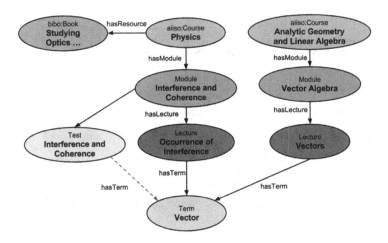

Fig. 3. An example of interdisciplinary relations between courses

Table 1. The results of ontology mapping between ontology of educational resources and TEACH ontology

Ontology of educational resources	Teaching Core Vocabulary
Classes	
aiiso:Course	teach:Course
Resource	teach:Material
Lecture	teach:Lecture
Properties	
hasTeacher	teach:teacher
isTeacherOf	teach:teacherOf

2.3   The Test Ontology

To describe the content of tests an upper ontology for test structure representation has been developed. Top-down approach was used to develop ontologies for the educational system because a new ontology extends existing upper ontology. The ontology[10] has the following classes: Test, Testing Knowledge Item, Group of Tasks, Task, Answer, Question, Fill-in the Blank, Matching, Multiple Choice, Single Answer, Text Answer, True/False. The ontology contains 12 classes, 10 object properties and 6 datatype properties. The classes of the developed ontology are shown in Fig. 4. The main purpose of the developed ontology is to represent structural units of a test and provide automatic task matching by defining semantic relations between tasks and terms [9].

The ontology has the class "Test" to store common test characteristics, e.g. its title and description, and class "Testing Knowledge Item" to describe test elements. The class "Testing Knowledge Item" has subclass "Task".

---

[10] http://purl.org/ailab/testontology.

Fig. 4. Main classes of the Test Ontology

The class "Group Of Tasks" [10] was added to group questions by parameters, e.g. by "difficulty". The class "Task" has the subclass "Answer". The class "Question" has subclasses describing question types: "Fill-in the Blank", "Matching", "Multiple Choice", "Single Answer", "Text Answer", and "True/-False". The class "Answer" has object properties "is wrong answer of" and "is right answer of". Using these two object properties instead of one data property "has answer" enables one to use one set of answers for many questions.

2.4   The Ontology of Student Activity in the E-Learning System

The ontology of student activity[11] is designed to store information about the student's learning process and results. Two upper ontologies have been used for its development: the Test Ontology, as described above, and the FOAF ontology[12] that describes people and relationships between them.

---

[11] http://purl.org/ailab/learningresults.
[12] http://www.foaf-project.org.

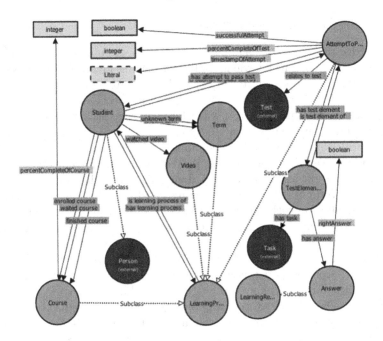

Fig. 5. Main classes of the student activity in the e-learning system ontology

The ontology contains 10 classes, 15 object properties and 5 datatype properties. The classes of the developed ontology are shown in Fig. 5. The class "Learning process" was added to store information about actions performed by a student in the system. Students can watch video (subclass "Video"), try to pass the test (subclass "AttemptToPassTest"), learn terms (subclass "Term") and pass a course (subclass "Course"). The ontology also has class "Student" to store information about users and their activity in system. This class is a subclass of class "Person" defined in FOAF ontology. The object properties "enrolled course", "finished course", and "subscribed course" describe relationships between the class "Student" and the class "Course". The class "Learning results" was added to store information about students educational activities and answers. Class "TestElement" contains information about "Task" (class of test ontology) and about student's "Answer" (subclass of class "LearningResults"), which can be correct or incorrect. Set of test elements constitutes attempt to pass test. The properties "timestamp of attempt" and "percent complete of test" allow e-learning system to store information about the time in which an attempt was made and to determine the result of the test. The e-learning system uses the ontology of tests and answers given by the user to build a list of terms that the user knows.

2.5   Ontology of Knowledge Rates

The knowledge rates ontology module is intended for keeping information about rates of term and domain knowledge rates for each student. The classes of the developed ontology are shown in the Fig. 6. Term's rate shows whether the student assimilated it. For example, if a student has watched or read the lecture with this term and has passed a test with this term successfully, we can consider, that student knows it. Ontology module contains class "Rate" and 5 subclasses: "Lecture Term Rate" computed as the number of lectures, containing this term and viewed by the student; "Test Attempt Term Rate" keeping attempts to pass a test with this term and number of correct answers to the task with a term; "Average Test Term Rate" based on average result of all attempts to pass one test or to pass all tests with this term; "Total Term Rate" based on sum of rates of this term; "Domain Rate" based on all rates of all terms from the domain student is learning. Each class contains data property "value" to store numeric values of rates. Also ontology contains object properties which link rates to the class "Student" from the educational ontology, "Test" from the test ontology and "Term" from the terms ontology. Ontology allows adding additional "Rate" subclasses storing new metrics as well as changing or adding formulas to calculate.

With the described modules we retain all the data associated with the training of students. Let us begin with a general example. John Smith, our imaginary student, has started the "Optics" course. This course contains lectures

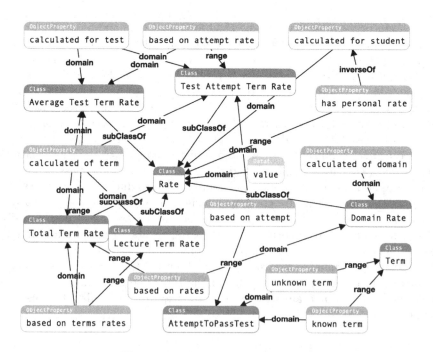

Fig. 6. Ontology of knowledge rates

and practical tasks (from ontology module of education), several tests with different groups of tasks and information about wrong and right answers (from ontology module of tests). Each task from tests and each lecture have terms which are described in ontology module of terms. When the student tries to pass the test a new individual of class "AttemptToPassTest" is created. When the student has solved the task his results are recorded in ontology module of knowledge rates. Based on wrong and right answers, metrics for terms that are checked in this tasks are compiled and changed, based on the formula, described below.

3 Methods of Ontology Population

3.1 Providers

The ECOLE system uses providers to collect educational linked open data. The provider supports automatic updating of the linked data from external resources. The provider can convert received structured data to RDF data model and store it into a triplestore. Each provider has separate context to manage the gathered data.

External Open Resources. The ECOLE system uses providers to get bibliographic data from open electronic libraries. The British National Bibliography (BNB)[13] provides open access to bibliographic content stored in RDF format. The bibliographic content is described using the BIBO ontology. The ECOLE system collects data about books and publications from BNB and allows tutors to link their courses with BNB content using "hasResource" property.

Linking Terms to DBpedia. The ECOLE system uses providers to get descriptions for terms of subject fields. Subject term in the system can be linked with external resources of knowledge bases, such as DBpedia, Freebase, and WikiData. The DBpedia provides a SPARQL-endpoint to content that was extracted from Wikipedia. The provider automatically collects description of the terms through queries to SPARQL-endpoint. This approach allows to expand subject term model using information from external resources.

3.2 Converting Structured Data to RDF

Many resources in the Web still stored in the structured format, but not in the RDF triples. For example, educational tests of university can be stored in XML format or electronic library can share content via a REST API. The ECOLE system uses conversion methods in the provider to integrate the structured data into the educational content.

---

[13] http://bnb.bl.uk/.

REST API. The ITMO University library shares information about books and papers via REST API. The ECOLE system uses groovy script to convert received data to RDF format. During conversion the system uses BIBO ontology to store data about books and papers in RDF format.

XML to RDF. To convert test data from XML format to RDF format a mapping was described. To provide conversion in the system an XMLProvider instance was created. The mapping for the test data conversion was described in the XML format. The mapping allows to convert XML files of the tests to the semantic data in accordance with the test ontology automatically. The XML-Provider uses XPath functions to extract data about objects and properties from the input XML file. The extracted data is converted into the RDF/XML format based on the mapping description.

An example of XML input, the mapping, and the output result for the test entity conversion is in Table 2.

3.3   NLP Algorithm

The developed NLP algorithm based on morpho-syntactic patterns is applied for terminology extraction from course tasks that allows to interlink the extracted facts (subject terms) to the instances of the systems ontology whenever some educational materials are changed. These links are used to gather statistics, evaluate quality of lectures' and tasks materials, analyse students answers to the tasks and detect difficult terminology of the course in general (for the teachers) and its understandability in particular (for every student).

Considering the small sample size and pre-set list of lecture terms POS-tag patterns combined with syntax patterns seem to be the most appropriate method for extracting terms from the tests [11–13]. The same algorithm was used for tests in Russian and for the tests translated into English for the demo version. About ten most typical compound term patterns were used to extract candidate terms (nominal compounds and adjectival compounds).

Russian compound candidate terms are transformed to the canonical form (that coincides with a headword in dictionaries) after extraction. For example, the pattern <adjective + noun>extracts an actual term <feminine adjective in instrumental case + feminine noun in instrumental case>, but lemmatization removes agreement and will produce two lemmas: <masculine adjective in nominative case>and <feminine noun in nominative case>whereas the appropriate form of the term is <feminine adjective in nominative case + feminine noun in nominative case>. This does not influence the procedure of linking candidate terms to the knowledge base instances but it is significant for the procedure of validation of missing terms.

NooJ linguistic engine [14] was used to extract terms. NooJ has a powerful regular expression corpus search allowing to join various POS-patterns in a single grammar to query the text. Dictionaries of lexical entries (for tests and ontology terms) and inflectional grammars were written for the Russian language

Table 2. Example of test entity conversion.

The input XML code

```
<test module="m_InterferenceAndCoherence"
     module_ns="Phisics"
     uri="TestOfInterferenceAndDiffractionFrenel"
     name="Test Of Interference And Diffraction Frenel">
</test>
```

The mapping code

```
<rule id="test"  nodeBase="//test"
     owlType="learningRu:Test"
     instanceNamespace="openeduTests"
     objectId="{./@uri}"
     objectLabel="{./@name}">
     <objectPropertyMapping nodeBase="."
          instanceNamespace="openeduTests"
          value="{./@name}"
          owlProperty="ifmotest:hasGroupOfTasks"
          referredRule="task_group" />
</rule>
```

The output RDF/XML code

```
<rdf:Description
     rdf:about="http://openedu.ifmo.ru/tests/
     TestOfInterferenceAndDiffractionFrenel">
  <rdf:type
     rdf:resource="http://www.semanticweb.org/
     k0shk/ontologies/2013/5/learning#Test"/>
  <label xmlns="http://www.w3.org/2000/01/rdf-schema#">
     Test Of Interference And Diffraction Frenel
  </label>
  <hasGroupOfTasks
     xmlns="http://www.semanticweb.org/
     fedulity/ontologies/2014/4/untitled-ontology-13#"
     rdf:resource="http://openedu.ifmo.ru/tests/
     Test_Of_Interference_And_Diffraction_Frenel"/>
</rdf:Description>
```

by the authors of the chapter. Lexical resources developed for the Russian language cover tasks' vocabulary totally. To analyze English text for the demo version standard NooJ resources were augmented and reused. NooJ dictionaries allow to combine various linguistic information for the lexical entry.

Several derivational paradigms for the Russian morphology were described with NooJ transducers and ascribed to the lexical entries [15]. Assigning derivational paradigms allows to produce a common lemma for the lexical entry and its derivatives, e.g. "coplanar" and "coplanarity" will have common lemma "coplanar". It should be noticed that NooJ descriptions allow to choose any word of the pair as a derivational basis and e.g. derive "coplanar" from "coplanarity" with a common lemma "coplanarity".

NooJ also has a very useful concept of a super-lemma. It allows for linking all lexical variants via a canonical form and store them in one equivalence class [16], e.g. in our dictionary a lexical entry "rectangular Cartesian coordinate system" is attached to its acronym "RCCS" (the last is a considered a canonical form) and a query either on acronym or on a compound term matches all the variants.

The overall algorithm of term extraction inside the NLP module consists of the following steps:

- Loading input in plain text into NooJ which performs its linguistic analysis using provided dictionaries. The output is also in plain text but with annotations containing morphological and semantic information for every analyzed word (Text Annotation Structure).
- Applying queries (that is POS-tag patterns combined with syntactic patterns) stored in a single NooJ grammar file to the Text Annotation Structure. The output is a list of candidate terms.
- The candidate terms with annotations are exported to a text file.

To apply the NLP-algorithm to other domains and languages one needs to compile NooJ lexical resources (dictionaries), write grammars and work out the templates to extract terms.

To map a candidate term to the system term via lemma, system terms were also lemmatized. Each system term has been assigned a text property "lemma" with a label containing the lemma of a term.

To handle links between system terms and test tasks the new data provider was implemented. The provider supports periodic updating of links. The input of the provider is the URI of the course entity. The provider handles all links between subject terms and test tasks of the input course.

The provider is based on the following algorithm:

- the provider collects tasks of the course using SPARQL queries;
- the provider forms the plain text content for each task using the information about questions and answers of the task;
- the provider launches NLP procedures in NooJ for the plain text content of the task;
- the provider extracts candidate terms from the NooJ output file, the data contain a canonical form and lemma(s) for the candidate term;
- the provider searches terms in the system to link them to candidate terms by using SPARQL queries; system terms and candidate terms are linked if they have the same lemma(s);
- the provider creates a link between selected system terms and the task by using the "hasTerm" property.

The algorithm of NLP provider is shown in Fig. 7.

If a word sequence extracted with a morpho-syntactic pattern does not match any of the system terms, it becomes a candidate instance to be included in the system as a new system term. It is also necessary to validate it, e.g., via external sources. We have chosen DBpedia to validate candidate instances.

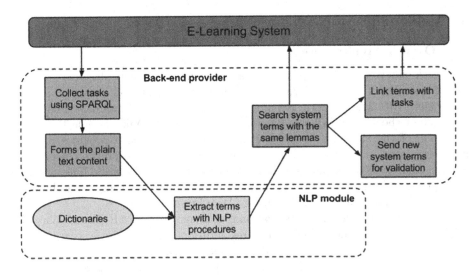

Fig. 7. The algorithm of NLP provider

A candidate instance is considered a new system term if its canonical form (a headword) completely matches the DBpedia instance's property "rdfs:label" or "dbpedia-owl:wikiPageRedirects", otherwise it is considered a false candidate. To avoid false matches results are filtered by the property "dcterms:subject". Validation is considered successful in case one or more DBpedia instances were matched. The validated candidate instance is added to the system as a new candidate term and is linked to the task. It becomes an authentic system term after teachers' approval.

Below is an example of a SPARQL query posted to the DBpedia's SPARQL-endpoint:

```
SELECT DISTINCT ?term {
    ?term dcterms:subject ?subject .
    VALUES ?subject {
        category:Concepts_in_physics
        category:Physical_optics
        category:Optics}
    {?name_uri dbpedia-owl:wikiPageRedirects ?term ;
        rdfs:label ?label .
    }
    UNION
    { ?term rdfs:label ?label }
    FILTER( STR(?label) ="Diffraction")
}
```

4   Implementation

4.1   Overall Architecture

The back-end is built on top of the Information Workbench platform[14]. The Information Workbench platform provides functions for interaction with Linked Open Data [16]. The platform is built on top of open source modules. The user interface of the system is based on the Semantic MediaWiki module [17]. An extension of the standard Wiki view Information Workbench provides predefined templates and widgets. RDF data management is based on the OpenRDF Sesame framework. The platform has support of SPARQL queries. The system has open SPARQL-endpoint for sharing its content. The front-end is implemented in Python[15] and uses the Django Web Framework[16].

The front-end collects the educational content from SPARQL-endpoint of the back-end. The front-end system stores additional data of the system, user's private data and user management data in local storage.

The overall architecture of the ECOLE system is shown in Fig. 8.

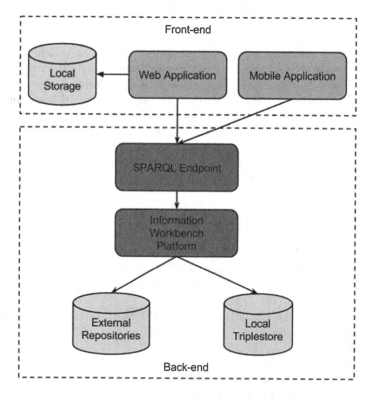

Fig. 8. The overall architecture of the ECOLE system

---

[14] http://www.fluidops.com/information-workbench/.

[15] https://www.python.org/.

[16] https://www.djangoproject.com/.

4.2   User Interface

The front-end of the ontology-based e-learning system is a lightweight Learning Management System. The front-end is designed to represent educational content conveniently. It also manages user data, user settings, and the results of the user's educational process. The front-end handles content administration, restricts access to educational content, and supports widgets for video lectures, tests, and practices.

The user interface of the front-end test page is shown in Fig. 9.

Data from the educational content are extracted with SPARQL queries to the Information Workbench SPARQL-endpoint [18]. The SPARQLWrapper Library[17] is used to compile SPARQL queries. When the system user has finished the test, the module gathering user's statistics sends the SPARQL Update Query [19] with user answers to the SPARQL-endpoint. When user statistics is gathered, objects having the type "AttemptToPassTest" and user's answers to the test's tasks are created in the system. The object with type "AttemptToPassTest" is bound to hash data of user's e-mail.

Upon completion of the test the information about amount of correct answers is displayed to the student. Also the list of subject terms for repeating is presented to the student. The system generates the list of problematic terms for the student using test results and relations between subject terms and tasks of the test. For each subject term of the test system counts the rank based on student's

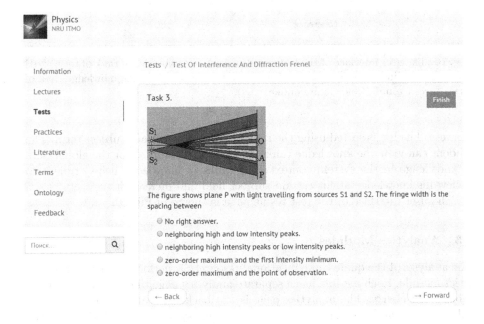

Fig. 9. The user interface of the front-end test page

---

[17] http://rdflib.github.io/sparqlwrapper/.

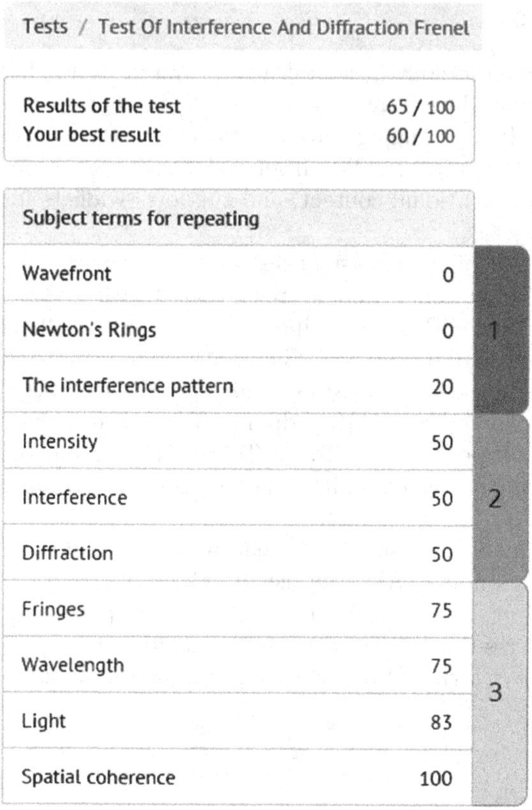

Tests / Test Of Interference And Diffraction Frenel

| Results of the test | 65 / 100 |
| Your best result | 60 / 100 |

Subject terms for repeating		
Wavefront	0	
Newton's Rings	0	1
The interference pattern	20	
Intensity	50	
Interference	50	2
Diffraction	50	
Fringes	75	
Wavelength	75	
Light	83	3
Spatial coherence	100	

Fig. 10. The user interface of the test result page. 1 - low knowledge rank of term, "red zone", 2 - medium knowledge rank of term, "yellow zone", 3 - high knowledge rank of term, "green zone" (Color figure online)

answers. The list is sorted using the rank of knowledge of each subject term. The student can view the knowledge rank of terms for certain test or the global rank of knowledge in the system context. The ranks of knowledge helps student to review his knowledge about certain subject field and increase it.

The user interface of the test result page is shown in Fig. 10.

4.3   Analytics Modules

The analysis of the quality of educational resources is performed inside the Analytics module. Each module has a separate analytics page that contains widgets, tables, and charts. The analytics page is the back-end system wiki page. The wiki page is based on the Semantic MediaWiki syntax and stored inside the Information Workbench system. The data of all UI elements on the page are obtained by using SPARQL queries.

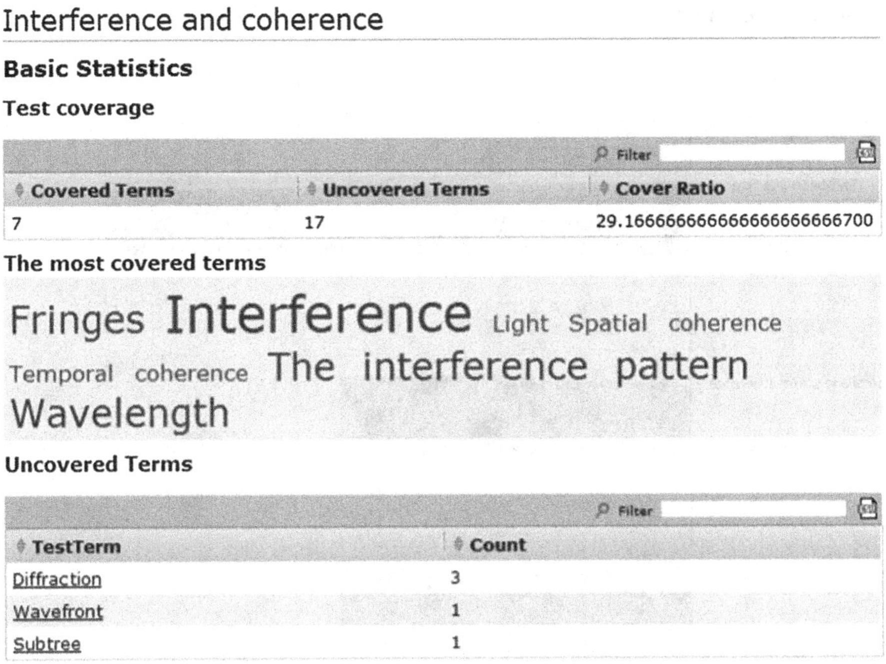

Interference and coherence

Basic Statistics

Test coverage

Covered Terms	Uncovered Terms	Cover Ratio
7	17	29.16666666666666666666700

The most covered terms

Fringes **Interference** Light Spatial coherence
Temporal coherence The **interference pattern**
Wavelength

Uncovered Terms

TestTerm	Count
Diffraction	3
Wavefront	1
Subtree	1

Fig. 11. User interface of the basic statistics

Lecture Coverage. The analysis of lecture coverage by tests is performed inside this module. Both test and lecture entities are associated with the module entity so the results can be obtained by analyzing terms of the test as well as terms of the lecture [20]. Each module has a separate analytics page that contains widgets, tables and carts. The data of all UI elements on the page are obtained by using SPARQL queries. The system analytics page of the module includes basic statistics and lecture coverage statistics. The basic statistics comprises:

- information about the total number of covered and uncovered terms of the module,
- cover ratio of the module based on the ratio of the number of covered terms among total number of module terms,
- a tag cloud of the most covered terms of the module,
- a table of the test terms not covered by lectures of the module.

User interface of the basic statistics is shown in Fig. 11.

The lecture coverage statistics shows the ratio of covered terms to the total number of lecture terms for each lecture of the module. The lecture coverage statistics is represented in a bar chart.

The user interface of the lecture coverage statistics is shown in Fig. 12.

Covering Lectures

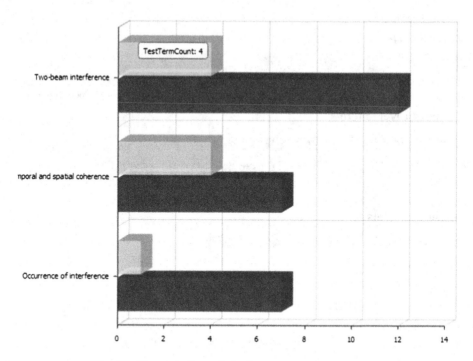

Fig. 12. User interface of the lecture coverage statistics

Troublesome terms. The data about the number of correct/incorrect user's answers allows the system to calculate the knowledge rating for any of the system terms. Using this rating, teachers can determine which terms of the course students know worst of all [21]. The knowledge rating is calculated by subtracting the number of incorrect answers from the number of correct answers for all tasks which contains this term. This metric is quite simple and could be replaced by a ranking formula after elaborating a set of features.

Data about user's answers is collected with the following SPARQL-query:

```
SELECT ?term
    (count(?correct_answer) AS ?correct_answer_count)
    (count(?answer) AS ?answer_count)
    ((2*?correct_answer_count - ?answer_count)
        AS ?rank)
WHERE{
    ?module learningRu:hasTest ?test   .
    ?test ifmotest:hasGroupOfTasks
        ?group_of_tasks .
    ?group_of_tasks ifmotest:hasTask ?task .
    ?test_element lres:hasTask ?task .
```

```
?test_element  lres:hasAnswer  ?answer  .
?task  learningRu:hasTerm  ?term  .
OPTIONAL {
     ?task  ifmotest:hasCorrectAnswer
          ?correct_answer
     filter ( ?correct_answer = ?answer)
}
}
GROUP BY ?term
ORDER BY ASC(?rank)
```

The analysis of troublesome terms in tests is performed inside the module. The system analytics page of the module includes a bar chart of the five most difficult terms for students and a table of all terms in the module with the knowledge rating. The user interface of the troublesome terms statistics is shown in Fig. 13.

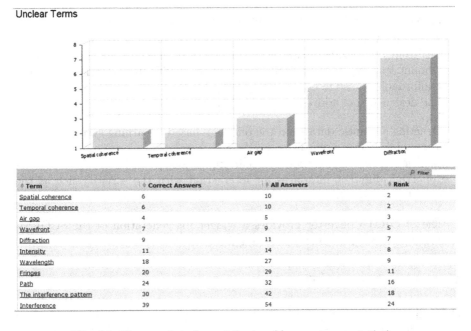

Fig. 13. The user interface of the troublesome terms statistics

The obtained set of analytical data helps evaluate adequacy of educational content and get a notion what content is ample or needs to be changed or added.

4.4 SCORM Converter

The learning content of the ECOLE system can not be integrated into Learning Management Systems such as Moodle. This is one of the main problems

Table 3. The number of objects in the dataset of ECOLE system)

Courses	4
Modules	14
Lectures	90
Tests	3
Subject fields	8
Subject terms	587
Subject terms mapped with DBpedia	219
Books	4094
Books from ITMO University	1131
Books from BNB	2951

of integrating systems learning content into local environment of the university. For this purpose the converter to the Sharable Content Object Reference Model (SCORM) was developed. The SCORM is a set of standards for e-learning systems [22,23]. The main goal of the SCORM Converter is to export learning courses from the ontology-based e-learning system and convert it to SCORM-conformant learning content. Exporting courses to SCORM standard makes it more affordable to use in e-learning systems.

The designed tool addresses a range of issues, such as:

– extracting semantic data from the ontology-based e-learning system,
– creating learning content using predefined templates,
– constructing SCORM-conformant learning content,
– supporting different interfaces for interaction with the service, such as the user interface and REST API.

The SCORM Converter provides a widget for generation of the SCORM-conformant learning content for certain course [24].

5 Evaluation and Results

The dataset of ECOLE system was created using manually obtained data. The part of the data was created by ontology population algorithms described in Sect. 3. The dataset contains objects of education such as courses, modules, lectures, subject fields and books. The statistics of the dataset of ECOLE system is shown in Table 3

Procedures for candidate term extraction and validation that were described in Sect. 3.3 have produced results displayed in Table 4. Term validation via DBpedia as it is proposed in the chapter is merely an idea rather than a technique. Using it in the described straightforward manner, we pursued the aim to remove the bulk of false candidates, not to validate the largest possible number

Table 4. Evaluation of the NLP-module (for the English language)

Percent of linked tasks, %	95
Percent of non-linked tasks, %	5
Number of different candidate terms	155
Number of manually extracted different terms	30
Percent of system terms, matched by candidate terms, %	50
Percent of candidate terms, matched by system terms, %	8
Percent of candidate terms to be included to the system terms after the validation procedure, %	6
Percent of false candidates, %	86

of the candidate terms. Overall 30 different terms were extracted manually from 20 tasks and 5 times more candidate terms were extracted using POS-patterns. The system contains 24 terms on interference and diffraction in the physics course at the moment.

The obtained results seem rather ambivalent: on the one side, 95 % of tasks were linked to at least one term. The leading system term is "Light", that has been linked to 12 tasks. On the other hand 50 % of system terms remained unlinked to tasks and about 60 % of them demand addition of proper tasks. The validation procedure using DBpedia as an external source provided 9 terms to be added as candidate system instances ("wavelength", "coherence", "coherent light", "diffraction", "amplitude", "aperture", "diffraction pattern", "optical path", "optical path length"). We treat all the remaining terms (that do not match any system term and failed DBpedia validation procedure) as false candidates. However, actually a few of false candidates are true terms that are not present in the chosen categories of DBpedia (Concepts_in_Physics, Optics and Physical_optics), but are present in other DBpedia categories (e.g. "Fresnel diffraction", "Fresnel zone" and "Fresnel number" are in the category: "Diffraction", "Michelson interferometer" is in the category "Interferometers"). Some terms have different canonical form in Russian and English, e.g. "Young's interference experiment" (is in DBpedia) corresponds to "Young's experiment" in Russian (no term in DBpedia). Thus, developers depend completely upon the data segmentation of the external source. Besides, there is a far more challenging problem: a task may not contain explicitly the term it is intended to check. Consider the following example:

A ladder is 5m long. How far from the base of a wall should it be placed if it is to reach 4m up the wall? Give your answer in metres correct to 1 decimal place.

This task checks understanding of the Pythagorean Theorem but it contains no explicit information allowing to assign proper keywords to the task. Such tasks are quite numerous.

Right now the algorithm fails to process such tasks leaving them unlinked. Elaborating the algorithm to handle cases like this is a part of future work.

6   Conclusion and Future Work

The developed ontologies and population methods for ECOLE system allow teachers to use various educational content from different external resources in their electronic courses. The developed modules for the system provide teachers with a tool to maintain relevance and high quality of existing knowledge assessment modules. With these modules tutors could fluently change education resources, content, and tests keeping them up-to-date. The ECOLE system provides rating of the terms which caused difficulties for students. Based on this rating teachers can change theoretical material of the course by improving description of certain terms and add proper tasks. The rating of the subject terms is also provided for the students. The rating helps students to find their knowledge gaps in subject fields and fill them.

The ECOLE system collects educational content from different sources and shares it with university learning systems. With ECOLE system exchange of the educational content between universities and other organizations can be implemented.

Future work of ECOLE system implies an integration of various data sources. Knowledge bases, such as Wikidata, can be integrated into the system to describe subject terms. Taxonomies of subject fields can be used for analysis of the relations between subject terms. These relations strengthen the importance of subject term.

Future work for the NLP-module implies describing a set of term periphrases. The algorithm should also filter out candidate terms that are non-thematic to the course, e.g. if a term "vector" occurs in a task on physics, it should not be marked as a term highly relevant to the course on interference because it is introduced in another course. The idea is that a link is created between a system term and any term that occurred in the task, but terms that do not belong to the topic of the course should not be marked as terms missing in the course.

Term extraction procedure can be also improved for adding parallel texts of tasks. The provider needs to be refined to create test entities in several languages.

The term knowledge rating can be also refined after its replacement by the proper ranking formula. The rating should be calculated using data about importance of the subject terms and user activity in the learning process.

The front-end of the ECOLE system can be found at

http://ecole.ifmo.ru

The source code of the developed ontologies can be found at

https://github.com/ailabitmo/linked-learning-datasets

The source code of the providers can be found at

https://github.com/ailabitmo/linked-learning-solution

Example of analytics for module"Interference and Coherence" can be found at

http://openedu.ifmo.ru:8888/resource/Phisics:m_InterferenceAndCoherence? analytic=1

The source code of the SCORM Converter can be found at

https://github.com/ailabitmo/linked-learning-scorm-converter

Acknowledgements. This work was partially financially supported by the Government of Russian Federation, Grant 074-U01.

References

1. Hendrix, M., Protopsaltis, A., Dunwell, I., de Freitas, S., Petridis, P., Arnab, S., Dovrolis, N., Kaldoudi, E., Taibi, D., Dietze, S., Mitsopoulou, E., Spachos, D., Bamidis, P.: Technical evaluation of The mEducator 3.0 linked data-based environment for sharing medical educational resources. In: The 2nd International Workshop on Learning and Education with the Web of Data at the World Wide Web Conference, Lyon, France (2012)
2. Khalili, A., Auer, S., Tarasowa, D., Ermilov, I.: Slidewiki: elicitation and sharing of corporate knowledge using presentations. In: ten Teije, A., Völker, J., Handschuh, S., Stuckenschmidt, H., d'Acquin, M., Nikolov, A., Aussenac-Gilles, N., Hernandez, N. (eds.) EKAW 2012. LNCS, vol. 7603, pp. 302–316. Springer, Heidelberg (2012)
3. Zablith, F.: Interconnecting and enriching higher education programs using linked data. In: Proceedings of the 24th International Conference on World Wide Web Companion. International World Wide Web Conferences Steering Committee (2015)
4. Mouromtsev, D., Kozlov, F., Parkhimovich, O., Zelenina, M.: Development of an ontology-based E-learning system. In: Klinov, P., Mouromtsev, D. (eds.) KESW 2013. CCIS, vol. 394, pp. 273–280. Springer, Heidelberg (2013)
5. Keler, C., d'Aquin, M., Dietze, S.: Linked data for science and education. Semant. Web 4(1), 1–2 (2013)
6. Kauppinen, T., Trame, J., Westermann, A.: Teaching core vocabulary specification. LinkedScience. org, Technical Report (2012)
7. McGuinness, D.L., Van Harmelen, F.: OWL web ontology language overview. W3C recommendation 10.10. (2004)
8. Halpin, H., Hayes, P.J.: When owl: sameas isn't the same: an analysis of identity links on the semantic web. In: LDOW (2010)
9. Soldatova, L., Mizoguchi, R.: Ontology of test. In: Proceedings of the Computers and Advanced Technology in Education, pp. 173–180 (2003)
10. Vas, R.: Educational ontology and knowledge testing. Electron. J. Knowl. Manage. 5(1), 123–130 (2007)
11. Hulth, A.: Improved automatic keyword extraction given more linguistic knowledge. In: Proceedings of the 2003 Conference on Empirical Methods in Natural Language Processing (EMNLP 2003), PP. 216–223 (2003)
12. Khokhlova, M.V.: Lexico-syntactic patterns as a tool for extracting lexis of a specialized knowledge domain. In: Proceedings of the Annual International Conference Dialogue (2012). (in Russian)

13. Bolshakova, E., Vasilieva, N.: Formalizacija leksiko-sintaksicheskoj informacii dlja raspoznavanija reguljarnyh konstrukcij estestvennogo jazyka [Formalizing lexico-syntactic information to extract natural language patterns]. Programmnye produkty i sistemy [Software and Systems], vol. 4, pp. 103–106 (2008)
14. Silberztein, M.: NooJ for NLP: a linguistic development environment (2002). http://www.NooJ4nlp.net/pages/NooJ.html
15. Silberztein, M.: NooJManual [Electronic resource], p. 99 (2003). http://www.NooJ4nlp.net/NooJManual.pdf
16. Haase, P., Schmidt, M., Schwarte, A.: The information workbench as a self-service platform for linked data applications. In: COLD (2011)
17. Krötzsch, M., Vrandečić, D., Völkel, M.: Semantic mediawiki. In: Cruz, I., Decker, S., Allemang, D., Preist, C., Schwabe, D., Mika, P., Uschold, M., Aroyo, L.M. (eds.) ISWC 2006. LNCS, vol. 4273, pp. 935–942. Springer, Heidelberg (2006)
18. Holovaty, A., Kaplan-Moss, J.: The Definitive Guide to Django. Apress, Berkley (2009). Estados Unidos: Editorial
19. Gearon, P., Passant, A. Polleres, A.: SPARQL 1.1 Update. World Wide Web Consortium (2013)
20. Parkhimovich, O., Mouromtsev, D., Kovrigina, L., Kozlov, F.: Linking E-learning ontology concepts with NLP algorithms. In: Proceedings of the 16th Conference of Open Innovations Association FRUCT (2014)
21. Kovriguina, L., Mouromtsev, D., Kozlov, F., Parkhimovich, O.A.: Combined method for E-learning ontology population based on NLP and user activity analysis. In: CEUR-WS Proceedings, vol. 1254, pp. 1–16 (2014)
22. Bohl, O., Scheuhase, J., Sengler, R., Winand, U.: The sharable content object reference model (SCORM)-a critical review. In: Computers in Education, pp. 950–951 (2002)
23. Qu, C., Nejdl, W.: Towards interoperability, reusability of learning resources: A SCORM-conformant courseware for computer science education. In: 2nd IEEE International Conference on Advanced Learning Technologies, Kazan, Tatarstan, Russia (2002)
24. Kozlov, F.: A tool to convert linked data of E-learning system to the SCORM standard. In: Klinov, P., Mouromtsev, D. (eds.) Knowledge Engineering and the Semantic Web. CCIS, vol. 468, pp. 229–236. Springer, Heidelberg (2014)

Use of Semantic Web Technologies in the Architecture of the BBC Education Online Pages

Eleni Mikroyannidi[✉], Dong Liu, and Robert Lee

British Broadcasting Corporation, Future Media Knowledge and Learning,
Salford, UK
{eleni.mikroyannidi,dong.liu,robert.lee}@bbc.co.uk

Abstract. The BBC has a rich collection of learning resources. The Knowledge and Learning Bitesize website aims to unlock the learning potential of this content for its users. To this purpose, the system employs semantic web technologies to organise the available learning resources. In this paper we describe the core data model that underlies the Knowledge and Learning website Beta (http://www.bbc.co.uk/education). The Curriculum Ontology formally describes the UK national curricula to allow users to easily discover content. We explain how the curriculum ontology supports the new version of BBC Knowledge and Learning website and discuss the challenges and benefits that such an architecture provides.

1 Introduction

Online learning resources have the potential to support a wide range of users. Each learner is an individual, with his or her own motivation for studying, accessing resources, and study habits and practices [5].

The BBC has understood the value of online learning from the early stages of the web, and has provided rich educational material to those wanting to learn. An example of this is the BBC Bitesize website[1], which started back in 1998 and is a popular formal education resource.

In the formal learning space the BBC has a number of sites: the already mentioned Bitesize, Skillswise[2] and LearningZone amongst others. There are tens of thousands of content items across these sites, with each site having different mechanisms for publishing, discovering and describing the content it serves.

To provide a coherent learning experience to users, a model for describing content in the context of the UK national curricula was developed. This model provided the foundation for building the new Knowledge & Learning website, presenting learning resources in the context of the UK national curricula in a consistent way. In addition, it allows for consistent reflection of changes in the national curricula throughout the product. Thus, the extensibility of the model is also an important feature.

---

[1] http://www.bbc.co.uk/bitesize/.

[2] http://www.bbc.co.uk/skillswise.

© Springer International Publishing Switzerland 2016
D. Mouromtsev and M. d'Aquin (Eds.): Open Data for Education, LNCS 9500, pp. 67–85, 2016.
DOI: 10.1007/978-3-319-30493-9_4

Designing the architecture of such a system is a challenging task. Each of the existing sites have similar yet different ways of describing and navigating through their content. In addition, the existing learning sites do not have a single content description model that could be reused in the site. Having a flexible structure in the back-end that can reflect the national curriculum and that can be used for consistently describing and organising learning resources is a key feature of the architecture.

We present the architecture behind the new Knowledge & Learning website and we focus on the curriculum ontology, which is central to the architecture. We show how it is used to describe and organise learning resources, how it supports navigation and how it is aligned with semantic markup vocabularies for Search Engine Optimisation (SEO). We will also present some of the challenges we faced and discuss future work.

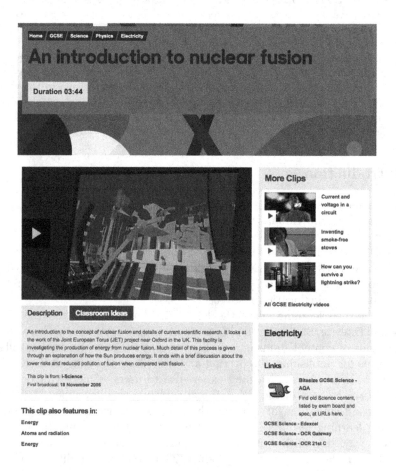

Fig. 1. An example video clip giving an introduction to nuclear fusion, for GCSE Physics. Taken from http://www.bbc.co.uk/education/clips/z4nwmp3.

2   The BBC Knowledge & Learning Online Content

The Knowledge & Learning Beta website aims to bring together the variety of BBC factual and learning content into a coherent model. At the time of writing, the education pages serve two types of learning content; (1) Video Clips and (2) Learner Guides.

Video Clips. Figure 1 shows an example video clip from the education pages. Video clips are related to a learning topic and are accompanied by classroom notes, which are notes on how a clip can be used in the classroom. The users for these clips are mainly teachers. The material provided in these pages aims to educate as well as stimulate the mind around topics in genres like history, science, arts. A video clip can be suitable for many topics of study, however, the classroom notes add value by providing information on the context of a programme of study. For example, the video clip of Fig. 1 is also featured in other topics of study like 'Energy' and 'Atoms and radiation'. However, the classroom notes can be different as the clip is presented in a different context.

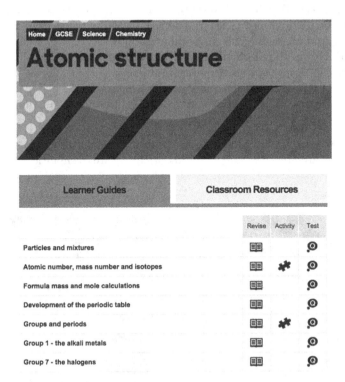

Fig. 2. Example learner guides. Taken from http://www.bbc.co.uk/education/topics/z84k7ty.

Learner Guides. Learner Guides are a rich interactive content format consisting of revision chapters, tests and activities. An example list of learner guides is shown in Fig. 2.

Learner Guides have a different learning purpose from video clips. They provide learning material in a stepwise way so that the learner moves from broader to more specific knowledge on a topic. Revision chapters provide an overview of educational material for a particular topic of study. Activities provide interactive material such as videos, games etc. The tests have multiple choice questions based on the revision chapters.

Both video clips and learner guides are structured based on the UK National Curricula. In the next section, we present the back-end architecture of this system.

3   Architecture Overview

Figure 3 shows the architecture behind the Knowledge & Learning Online Pages. The main actors in this architecture are:

- **Curriculum Ontology:** The ontology is used for describing the curriculum vocabulary.
- **Curriculum API:** The API used to query the Linked Data Platform.
- **Linked Data Platform (LDP):** The BBC internal triple store and services for saving and managing the curriculum metadata.
- **Content Items like Learning Clip and Learner Guides:** The learning resources that are shown in the education pages.
- **Content API:** The API for querying the Content Store.
- **Content Store:** The system where learning resources are authored and stored.

The Knowledge & Learning uses a Dynamic Semantic Publishing (DSP) model in its architecture [13]. In this architecture, semantic web models

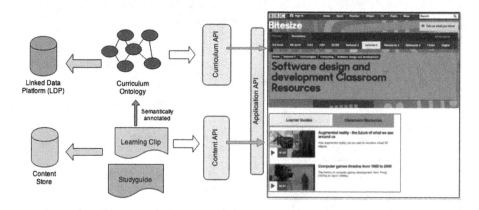

Fig. 3. The BBC knowledge & learning architecture.

and linked data are key features. Similarly, in the Knowledge & Learning Online Pages, different components of the front end are served by different systems in the back-end. This is depicted in Fig. 3. The curriculum ontology is the core component of the architecture, which supports the navigation and organises the learning resources based on the UK national curricula.

In a nutshell, the ontology and its instance data are served by the Linked Data Platform, which amongst other services it provides the BBC internal triple store. Learning content such as video clips and learner guides are saved as XML Documents in a different system named as Content Store. The Content Store serves all the educational resources that are shown in the online pages and it is also the main system used by the editorial team for authoring new content. The coupling between Learning resources and curriculum ontology is done through semantic annotation. In particular, the content items are tagged with curriculum instances so that the Application Layer can retrieve related content for curriculum ontology instances. The definition of new curriculum instances and the tagging procedure is part of the workflow for publishing content. Our editorial teams initially define the vocabulary (fields of study, topics of study etc.) that is used for annotating educational content. Then video clips and learner guides are tagged with the corresponding topic of study. In this way the Application Layer can retrieve the corresponding topics of study and their associated content and render this information in the online pages.

In the following sections, we will focus on the curriculum ontology and show the benefits of linked data for supporting online learning resources. One main benefit is that using linked data in the backend can offer flexibility on the aggregations of content. In addition we will present how the ontology is mapped with learning markup vocabularies and how linked data are used effectively with markup for improving search.

4   The Curriculum Ontology

The Curriculum Ontology is a core data model for formally describing the National Curricula across the UK. The full documentation as well as the latest version of the ontology are available online[3]. The instance data of the curriculum ontology are published on GitHub[4].

The Curriculum Ontology aims to:

- provide a model of the national curricula across the UK
- organise learning resources, e.g. video clips and learner guides
- allow users to discover content via the national curricula

The Curriculum Ontology has been designed to organise content in a way that allows students and teachers to navigate and discover learning resources. It achieves this by providing broad units of learning (e.g. a Topic) and more finely grained units (e.g. a Topic of Study). Figure 4 depicts the classes and properties in the curriculum ontology.

---

[3] http://www.bbc.co.uk/ontologies/curriculum.
[4] https://github.com/BBC/curriculum-data.

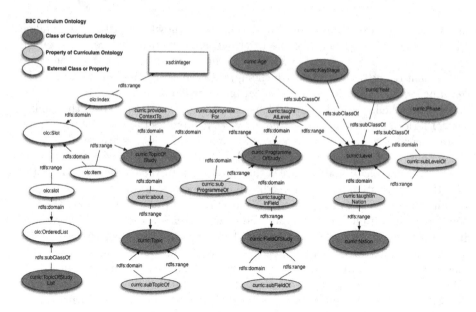

Fig. 4. The curriculum ontology diagram.

4.1   Core Concepts of the Ontology

The core concepts in the ontology are (1) Level, (2) Field of study, (3) Programme of Study, (4) Topic of Study and (5) Topic. In the ontology, these concepts are represented as OWL Classes [9]. The rationale for these concepts in the domain model is outlined below.

Level and Its Sub-concepts. Level refers to different stages of education. Typically, there are four different ways to specify the levels: Age Range, Year, Key Stage and Phase. Thus, in the ontology these concepts are modeled as subclasses of Level.

Key Stage is a way to specify the stage of the state education system in England, Wales, Northern Ireland. Some example Key Stages are KS3, GCSE etc. Year is a way to specify the stages of education.

The Phase sub-concept is borrowed from the controlled terms for describing phase of education, which are published by the data.gov.uk[5]. The phases of education include Primary, Secondary and 16-Plus.

Fields of Study. It refers to the discipline of a curriculum. Some example fields of study are Science, Maths, English Literature etc.

---

[5] http://education.data.gov.uk/.

Programmes of Study. A programme of study is the combination of a nation, an educational level and the subject (Field of Study) being studied. Thus, in the ontology, the Field of Study class is connected to the Field of Study with the taughtInField predicate and to the Level with the taughtAtLevel predicate. Some example programmes of study are 'GCSE Maths', 'Higher Biology' etc.

Topics of Study. A Topic of study is a topic within the context of a programme of study. It aims to provide a formal learning context to an asset or a collection of assets. An example topic of study is 'Circuits', which is taught in the 'KS2 Science' programme of study.

Ordering Topics of Study. The ordering of Topics of Study is a key requirement in the Curriculum Ontology, because some Topics of Study can be prerequisites of others. For instance, students have to learn the English alphabet before English grammar. The TopicOfStudyList class uses the external Ordered List Ontology [2] to curate the sequences of Topics of Study. This is achieved by assigning an indexed slot to each Topic of Study. For allowing multiple indexing per Topic of Study, instead of directly assigning the index in the Topic of Study, the TopicOfStudyList class is used, which is subclass of the OrderedList class from the external Ordered List Ontology. This pattern allows a Topic of Study to appear in many lists and have different order in each list. The Ordered List Ontology is also described in the Ontology Design Pattern (ODP) catalogue[6] [8].

Topics. A Topic can highlight the content of the learning resources in a more specific way than the Field of Study. For example, energy is a topic of physics.

Topics of Study Viewed as Topics. In the curriculum ontology, the difference between a Topic of study and a topic is that mapping content to a Topic of Study makes it easy to find specific content whilst mapping to a generic Topic allows users to discover a wider range of content.

A Topic of Study is defined as a Topic in the context of Programme of Study. It addresses the following issues:

- **A Topic Across Levels.** The meaning of a Topic, e.g. 'Geometry (Shape & Space)', can vary for different Levels. Geometry for KS1, usually called 'Shape & Space', needs a different description to Geometry in KS3 as KS1 is typically for primary-age students whereas KS3 is for secondary-age students.
- **A Topic Across Fields of Study.** For example, 'Energy Resources' is a Topic of both Physics and Geography, but the learning content for 'Energy Resources for Physics' and 'energy resources for geography' can be different.
- **Topic Synonyms.** The topic of 'algebra' is usually described as 'relationships' in the Scottish national curricula and as 'Algebra' in the English curricula.

---

[6] http://ontologydesignpatterns.org/wiki/Ontology:Ordered_List_Ontology.

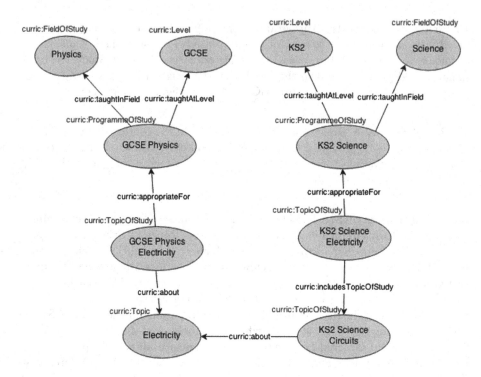

Fig. 5. Example instantiation of the the core concepts from the curriculum ontology.

Fields of Study Viewed as Topics. A Field of Study could also be viewed as just a Topic covering a broader area of learning. The reasons for defining them as two separate concepts are:

- **Fields of Study are published by Government bodies.** For instance, the Fields of Study taught at primary schools in England are published by the government here. In general the national curricula define the breadth of a field of study but do not provide a prescriptive list of individual topics, although in some levels (e.g. GCSE) exam boards do specify a list of Topics of Study for each Field of Study.
- **Usage of topics across different Fields of Study.** As mentioned previously, some Topics can be used across Fields of Study.

5   Describing Learning Content with the Curriculum Ontology

Effective description and organisation of learning content is achieved by semantically annotating learning content with instance data from the Curriculum Ontology. Figure 5 shows an example instantiation of the core concepts from the curriculum ontology.

Figure 5 depicts the interlinking of two topics of study ('GCSE Physics Electricity' and 'KS2 Science Circuits') that belong to different programmes of study, but they are around the same topic, which is 'Electricity'. The 'GCSE Physics' programme of study has a relationship with the corresponding Field ('Physics') and Level ('GCSE')[7]. The information about a topic of study holds also information about the level, programme of study and related topic. Having such a structure in the backend has multiple benefits, such as achieving more dynamic aggregations of content and providing a meaningful organisation of the content that reflects the UK National Curricula.

5.1   Dynamic Semantic Publishing

The Curriculum Ontology is the glue that holds the content together and the basis of the website navigation. Following a Dynamic Semantic Publishing approach we moved away from a relational publishing model to one that separates semantics from content and allows dynamic aggregations.

Figure 6 shows an example of the use of the curriculum ontology instance data for organising content and for the provision of the main navigation in the website.

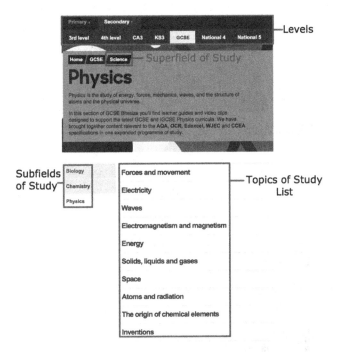

Fig. 6. Example showing the curriculum ontology supporting the BBC knowledge & learning online navigation.

---

[7] More precisely, 'GCSE' is the qualification (Phase) but it is inferred to be a Level too.

In particular, it shows how the topic of study **GCSE Physics Electricity** and related data of Fig. 5 are shown in the front end. Related information about the Level, Superfield (**Science**) and sibling fields (**Biology, Chemistry, Physics**) are presented in a hierarchical way in the navigation panel. This information is retrieved by querying the triple store through the Curriculum API. The curriculum API provides the result of the SPARQL queries in the Application API (Fig. 3). In addition, the list of topics of study shown in Fig. 6 is implemented with the Ordering List Ontology Design Pattern (ODP), described in Sect. 4.1.

The workflow for publishing new content in the website is shown in Fig. 7. Editorial teams create new content based on a DSP approach. The content that is commissioned is always in the context of the UK National Curricula. That allows the editorial team to initially define the curriculum vocabulary for the new content. These are the instance data of the curriculum ontology. An example is the creation of a new programme of study e.g. 'KS1 Computing' and topics of study for this programme. On the second step editorial team authors new content like study guides and learning clips. The DSP approach allows to associate the curriculum ontology instance data with the content. For example, a video for KS1 Computing is tagged with a corresponding topic of study from the curriculum ontology, which is named as 'Computer science'. Figure 8 shows how

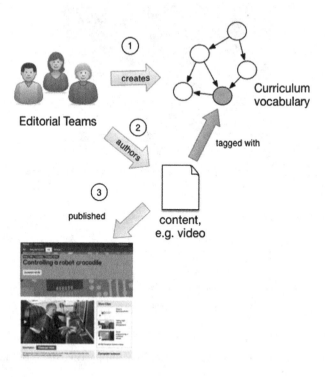

Fig. 7. Dynamic semantic publishing workflow for bitesize content.

Fig. 8. A semantically annotated video for bitesize in the content store.

tagging happens in the content store. A member from the editorial team, applies the curriculum topic of study 'Computer Science' to a learning clip about controlling a robot crocodile.

As mentioned in Sect. 3, the description of new content is saved in the form of XML in the content store. The web interface for creating new content contains a tag picker where the content is associated with a particular topic of study from the curriculum ontology data. To achieve this, the content store is connected with the LDP for picking topics of study. A consistent tagging approach is adopted, where content in the context of the Curriculum is always tagged with a topic of study. Thus in the web form of Fig. 8, the editorial team can select only a topic of study to associate it with the clip. There are two reasons for this. First, the content is coupled in the navigation with topics of study and second, topics of study are interlinked with other concepts from the curriculum ontology such as programme of study, field of study etc. so this information can be retrieved through property paths.

The use of Linked Data in the architecture of the system can help towards a dynamic aggregation of content. Additional views can be implemented by querying a different part of the graph like for example, creating an aggregation view of content grouped by topic. Figure 9 shows how a video can be used in different education context. For example, the figure shows that a learning clip about 'The social effects of automation' used for the topic of study, 'Computers in society' also features in other topics of study, such as 'the history of ICT' and 'Industrial and commercial applications'. Thus, a clip can be used in the context of multiple topics of study with different classroom ideas. In this way the user can browse other related topics to a particular clip.

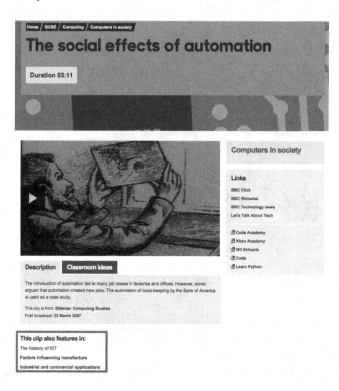

Fig. 9. The use of a learning clip in multiple topics of study.

5.2 SPARQL Support

The main views and navigation of the website are driven by the curriculum LDP data. In particular, for every view a SPARQL CONSTRUCT query is run for creating this view by querying the right part of the curriculum graph.

For example, the SPARQL CONSTRUCT query shown in Fig. 10 aims to retrieve the programmes of study associated with a specific educational level i.e. GCSE. The navigation components on the corresponding Web page[8] of the GCSE level are shown according the results of this query. The presented SPARQL query returns the label, description and the depiction image of the relevant programmes of study and fields of study, as well as detailed information about the required educational level itself. In addition, the results of this query also tells for which nation the educational level is appropriate.

As shown below, the SPARQL query was written using concepts and predicates defined in the aforementioned BBC Curriculum Ontology and the BBC Core Concepts Ontology[9]. In more detail, Line 13 to 18 of the SPARQL query gathers information about the given educational level, while Line 19 to 22 trying

---

[8] http://www.bbc.co.uk/education/levels/z98jmp3.
[9] http://www.bbc.co.uk/ontologies/coreconcepts.

```
1   CONSTRUCT {
2     ?klPos a curric:ProgrammeOfStudy ;
3       rdfs:label ?posLabel ; dc:description ?description ;
4       curric:taughtAtLevel ?klLevel ; curric:taughtInField ?klField ;
5       klal:depictionPID ?posDepictionPID ; curric:subProgrammeOf ?klSuperPos .
6     ?klLevel a ?levelType ; curric:taughtInNation ?nation ;
7       rdfs:label ?levelLabel ; dc:description ?levelDescription .
8     ?klField a curric:FieldOfStudy ; rdfs:label ?fieldLabel ;
9       klal:depictionPID ?fieldDepictionPID ; dc:description ?fieldDescription ;
10      curric:subFieldOf ?klSuperField .
11    ?nation a curric:Nation ; rdfs:label ?nationLabel .
12  } WHERE {
13    ?level a ?levelType ;
14      core:sameAs <http://www.bbc.co.uk/education/levels/z98jmp3#level> ;
15      core:sameAs ?klLevel ; core:preferredLabel ?levelLabel .
16    OPTIONAL {
17      ?level curric:preferredDescription ?levelDescription .
18    }
19    OPTIONAL {
20      ?level curric:taughtInNation ?nation .
21      ?nation a curric:Nation ; core:preferredLabel ?nationLabel .
22    }
23    OPTIONAL {
24      ?pos a curric:ProgrammeOfStudy ; core:sameAs ?klPos ;
25        core:preferredLabel ?posLabel ; klal:depictionPID ?posDepictionPID ;
26        curric:taughtAtLevel ?level ; curric:taughtInField ?field .
27      ?level core:sameAs <http://www.bbc.co.uk/education/levels/z98jmp3#level> .
28      OPTIONAL {
29        ?pos curric:preferredDescription ?description .
30        OPTIONAL {
31          ?pos dc:description ?descriptionlang
32            BIND(?descriptionlang as ?description)
33            FILTER ( lang(?descriptionlang) = "en-gb") .
34        }
35      }
36      OPTIONAL {
37        ?pos curric:subProgrammeOf ?superPos .
38      }
39      ?field a curric:FieldOfStudy ; core:sameAs ?klField ;
40        klal:depictionPID ?fieldDepictionPID ; rdfs:label ?fieldLabel
41        FILTER ( lang(?fieldLabel) = lang(?levelLabel) ) .
42      OPTIONAL {
43        ?field curric:preferredDescription ?fieldDescription .
44      }
45      OPTIONAL {
46        ?field curric:subFieldOf ?superField .
47        ?superField core:sameAs ?klSuperField .
48      }
49    }
50  } ORDER BY ?fieldLabel
```

Fig. 10. The SPARQL query for retrieving information about the educational level GCSE.

to get the nations in which the education level is taught. Line 23 to 38 generates a set of the programmes of study which are linked to the given level via *curric:taughtAtLevel*. Furthermore, for the purpose of rendering the Web page, we use Line 39 to 48 of the query to retrieve the fields of study associated with those programmes of study.

Figure 11 shows the RDF triples aggregated by executing the SPARQL query against the data repository. Those RDF statements describe the programmes of study related to the educational level of GCSE. It is worth noting that the resulting RDF triples forms a graph since we use CONSTRUCT to query the RDF repository. The reason for using CONSTRUCT instead of SELECT is that it facilitate parsing the results, because all the data are in the format of RDF.

```
<http://www.bbc.co.uk/education/levels/z98jmp3#level> a curric:KeyStage, curric:Level;
    rdfs:label "GCSE"@en-gb ;
    dc:description "GCSE is the qualification taken by 15 and 16 year olds to mark their graduation from the Key Stage 4 phase
of secondary education in England, Northern Ireland and Wales."@en-gb ;
    curric:taughtInNation <http://www.bbc.co.uk/things/00eb010f-568a-4b89-bbfe-799d5b812bed#id>,
        <http://www.bbc.co.uk/things/06dbdeed-0f5e-41f7-b2ef-0f3e3039a72f#id>,
        <http://www.bbc.co.uk/things/9ba4d1e5-7cc7-48c0-9793-ad198403c54c#id> .

<http://www.bbc.co.uk/things/00eb010f-568a-4b89-bbfe-799d5b812bed#id> a curric:Nation ;
    rdfs:label "Wales"@en-gb .

<http://www.bbc.co.uk/things/06dbdeed-0f5e-41f7-b2ef-0f3e3039a72f#id> a curric:Nation ;
    rdfs:label "Northern Ireland"@en-gb .

<http://www.bbc.co.uk/things/9ba4d1e5-7cc7-48c0-9793-ad198403c54c#id> a curric:Nation ;
    rdfs:label "England"@en-gb .

<http://www.bbc.co.uk/education/subjects/zpsvr82#programme-of-study> a curric:ProgrammeOfStudy ;
    rdfs:label "GCSE Business"@en-gb ;
    dc:description "GCSE Business Studies is designed for students finishing secondary school to learn skills for running a
business, such as managing money, advertising and employing staff."@en-gb ;
    klal:depictionPID "p017b7l9" ;
    curric:taughtAtLevel <http://www.bbc.co.uk/education/levels/z98jmp3#level> ;
    curric:taughtInField <http://www.bbc.co.uk/education/subjects/zjnygk7#field-of-study> .

<http://www.bbc.co.uk/education/subjects/zrkw2hv#programme-of-study> a curric:ProgrammeOfStudy ;
    rdfs:label "GCSE Science"@en-gb ;
    dc:description "Science is the systematic study of the physical and natural world through observation and
experimentation."@en-gb ;
    klal:depictionPID "p017dpn3" ;
    curric:taughtAtLevel <http://www.bbc.co.uk/education/levels/z98jmp3#level> ;
    curric:taughtInField <http://www.bbc.co.uk/education/subjects/z7nygk7#field-of-study> .

<http://www.bbc.co.uk/education/subjects/z2f3cdm#field-of-study> a curric:FieldOfStudy ;
    rdfs:label "Geography"@en-gb ;
    dc:description "Geography is the study of the shape and features of the Earth's surface, including countries, vegetation,
climates and how humans use the world's resources."@en-gb ;
    klal:depictionPID "p017dmsh" .

<http://www.bbc.co.uk/education/subjects/zjnygk7#field-of-study> a curric:FieldOfStudy ;
    rdfs:label "Business"@en-gb ;
    dc:description "Business studies covers the different skills for running a business, such as managing money, advertising and
employing staff."@en-gb ;
    klal:depictionPID "p017b7l9" .

<http://www.bbc.co.uk/education/subjects/z7nygk7#field-of-study> a curric:FieldOfStudy ;
    rdfs:label "Science"@en-gb ;
    dc:description " "@en-gb ;
    klal:depictionPID "p017dpn3" .
```

Fig. 11. RDF triples generated by executing the SPARQL query.

5.3   Mapping to LRMI Vocabulary

The Learning Resource Metadata Initiative (LRMI) [1,15] has been adopted by http://schema.org and aims to establish an open standard for adding semantic mark-up to online learning resources. Using the LRMI vocabulary enables easier discovery of content by search engines and other organisations. Thus, mapping concepts from the BBC curriculum ontology to the LRMI vocabulary contributes towards a model of consistent organisation and discovery of content. The conceptual mappings between the Curriculum Ontology and LRMI are shown in the Fig. 12.

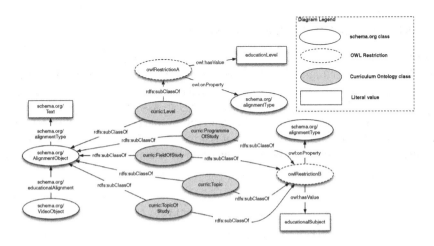

Fig. 12. A model representing the mappings of the curriculum ontology to Schema.org LRMI vocabulary.

Instead of modeling school curricula, LRMI provides a generic framework for describing learning resources, which is independent from certain educational frameworks. To this end, LRMI introduces the concept AlignementObject and the educationalAlignment property [3]. The AlignementObject is an abstract concept mapped to educational levels, subjects and topics. The educationalAlignment links a learning resource with an educational concept. The Curriculum Ontology classes share the same intent as the Schema.org AlignmentObject, thus they are defined as sub-concepts of the AlignmentObject concept.

The AlignmentObject class provides an alignmentType property that describes the type of alignment being specified. In Fig. 12, there are two types of alignment, the 'educationLevel' and the 'educationalSubject'. These types allow alignment to the corresponding Curriculum Ontology classes. OWL restrictions are used to enforce that correct alignmentType properties are used. Thus, if we want to say that all instances of curric:Level are of schemorg:AlignmentType "Educational-Level" this can be implemented in OWL as:

```
curric : Level a owl : Class ;
    rdfs : subClassOf _ : owlRestrctionA ;
    rdfs : subClassOf schemorg : AlignmentObject .

_ : owlRestrctionA a owl : Restriction ;
    owl : onProperty schemorg : alignmentType ;
    owl : allValueFrom " EdcuationalLevel " .
```

Similarly every instance of a curric:ProgrammeOfStudy, a curric:Topic, curric: FieldOfStudy, curric:TopicOfStudy can be implied to be an schemorg: AlignmentType "edcationalSubject".

5.4   Markup on BBC Education Website

In order to feed search engines with metadata about learning resources and the UK curricula, semantic markup is added to the HTML pages of a content item, i.e. a video clip. Figure 13 demonstrates an example markup using typicalAgeRange and educationalAlignment. GCSE is associated to an instance of VideoObject. The content of the alignmentType property indicates that GCSE is an educational level. Similarly, a field of study and topic of study can be defined as educational subjects by defining the value of alignmentType to educational-Subject.

```
<div itemscope="itemscope" itemtype="http://schema.org/
    VideoObject">
    ...
    <meta itemprop="typicalAgeRange" content="14-16" />
    <meta itemprop="educationalAlignment" itemscope=""
        itemtype="http://schema.org/AlignmentObject">
        <meta itemprop="targetName" content="GCSE" />
        <meta itemprop="targetUrl" content="http://www.bbc.co
            .uk/education/levels/z98jmp3" />
        <meta itemprop="targetDescription" content="GCSE is
            ..." />
        <meta itemprop="alignmentType" content="
            educationLevel" />
    </meta>
    ...
</div>
```

Fig. 13. Example of semantic markup.

Google Custom Search engines[10] can be easily built with the help of semantic markup. A custom search engine built with refinement for levels GCSE, KS3 (more:p:AlignmentObject-name:KS3, more:p:AlignmentObject-name:GCSE)[11]

---

[10] https://www.google.com/cse.

[11] https://www.google.com/cse/publicurl?cx=005635636900202455771:bttbeggy8g0.

and other levels is shown in Fig. 14. The screenshot shows the results when searching for 'Hamlet'. The results can be categorised by level, e.g. show video clips that are appropriate for level KS3.

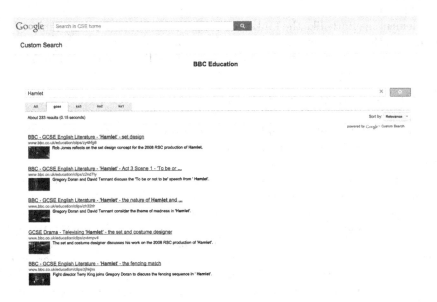

Fig. 14. Google custom search engine for BBC education.

6   Related Work

For this work a number of online education resources and vocabularies were researched. The UK government[12] provides a very useful structure in their online educational resources, which were considered in the curriculum ontology as well. In particular, the notion of Educational Phase is also used in the BBC curriculum ontology. Other ontologies like the Bolowgna ontology [7] and the ROLE Learning ontology [4] have been developed as part of educational projects before the curriculum ontology. The purpose of the Bolownga ontology was to model an academic setting and to support the publication and exchange of information among universities. The ROLE Learning ontology was developed to support self-regulating learning and it represents a Psycho-Pedagogical Integration Model of connecting learning strategies, techniques and activities.

In [11], MONTO, A Machine-Readable Ontology for Teaching Word Problems in Mathematics is described. The MONTO ontology is based on a Mathematical thinking framework for the representation of the users cognitive model and learning strategy and it is aligned with the generation of domain specific topics (e.g. learning the topic of Circumference in Mathematics).

---

[12] http://education.data.gov.uk/.

In addition, the work from [12,14] was also reviewed when designing the curriculum ontology. It is one of the most structured resources related to education that preexisted the curriculum ontology and its analysis was a precursor to the development of this ontology.

The support of education data with Semantic Web Technologies [6] has been a key element in efforts like the Linked Universities initiative[13] and the LinkedUp[14] project. In this work semantically annotate learning content using the curriculum ontology. The details of semantic annotation of media are presented in [10].

7 Discussion and Future Work

The BBC offers tens of thousands of learning resources across its sites, with each site having different mechanisms for publishing, discovering and describing the content it serves.

To improve consistency, we incorporated ontology models and linked data in the architecture of the BBC Knowledge & Learning Beta Online Pages. In particular, the development and use of the curriculum ontology in the architecture of the system allowed for interlinking curriculum concepts allowing every content item to be semantically annotated with relevant curriculum topics. In addition, content can be discovered consistently and shown in context with other similar content. This is achieved by semantically annotating learning content with Curriculum Ontology instance data. Mapping the Curriculum Ontology concepts with learning markup vocabularies, such as the Learning Resource Metadata Initiative (LRMI), allowed for better precision in search using the metadata of the learning content.

Hiding the model's complexity and providing a consistent navigation is a challenge of the architecture, which is achieved with the implementation of services and APIs on the top of the linked data. A benefit of the ontology supporting the BBC online education pages is that it can offer dynamic aggregations of content achieved by querying the linked curriculum data. It can also help to easily discover content. For example, the recommendations on other relevant topics to a clip is done via the ontology data. Building additional recommendation services using the curriculum ontology and other ontologies is a good future work direction.

The adoption of an ontology model in the architecture of the system also allows the seamless extension of the data to reflect changes in the National Curricula. For example there are plans for the inclusion of the '16 Plus' Phase and its corresponding Programmes of Study in the Bitesize pages.

One key requirement of the new Knowledge & Learning is to provide a consistent model reflecting the UK National curricula where users can learn more about science, nature, history, religion, arts and more, in a continuous learning journey. Semantic Web Technologies and Linked Data can give a leverage in accomplishing this task as they allow an effective interlinking and querying of web data.

---

[13] http://linkeduniversities.org/lu/.
[14] http://linkedup-project.eu/.

Acknowledgments. The authors would like to thank Ali Craigmile, Jemma Summerfield, Paul Rissen, Amaal Mohamed and Zoe Rose for their help on the design of the BBC Curriculum Ontology.

References

1. The Learning Resource Metadata Initiative (LRMI) Specification. http://www.lrmi.net/the-specification. Accessed 03 July 2014
2. Abdallah, S.A., Ferris, B.: The Ordered List Ontology Specification (2010). http://smiy.sourceforge.net/olo/spec/orderedlistontology.html. Accessed 03 July 2014
3. Barker, P.: Explaining the LRMI Alignment Object [blog post] (2014). http://blogs.cetis.ac.uk/philb/2014/03/06/explaining-the-lrmi-alignment-object/
4. Berthold, M., Dahn, I., Kiefel, A., Lachmann, P., Nussbaumer, A., Albert, D.: ROLE learning ontology: an approach to structure recommendations for self-regulated learning in personalized learning environments. In: The Future of Learning Innovations and Learning Quality, p. 104 (2012)
5. Collison, G., Elbaum, B., Haavind, S., Tinker, R.: Facilitating online learning: Effective strategies for moderators. In: ERIC (2000)
6. d'Aquin, M.: Linked Data for Open and Distance Learning (2012)
7. Demartini, G., Enchev, I., Gapany, J., Cudré-Mauroux, P.: The bowlogna ontology: fostering open curricula and agile knowledge bases for europe's higher education landscape. Semant. Web **4**(1), 53–63 (2013)
8. Gangemi, A., Presutti, V.: Ontology Design Patterns. In: Staab, S., Studer, R. (eds.) Handbook on Ontologies. International Handbooks on Information Systems, pp. 221–243. Springer, Heidelberg (2009)
9. Horrocks, I., Patel-Schneider, P.F., Van Harmelen, F.: From SHIQ and RDF to OWL: the making of a web ontology language. Web Semant. Sci., Serv. Agents World Wide Web **1**(1), 7–26 (2003)
10. Kobilarov, G., Scott, T., Raimond, Y., Oliver, S., Sizemore, C., Smethurst, M., Bizer, C., Lee, R.: Web-How, media meets semantic the BBC uses DBpedia, linked data to make connections. In: Aroyo, L., et al. (eds.) The Semant. Web: Res. Appl. LNCS, vol. 5554, pp. 723–737. Springer, Heidelberg (2009)
11. Lalingkar, A., Ramanathan, C., Ramani, S.: MONTO: A machine-readable ontology for teaching word problems in mathematics. J. Edu. Technol. Soc. **18**, 197–213 (2015)
12. Morris, R.D.: Web 3.0: Implications for online learning. TechTrends **55**(1), 42–46 (2011)
13. Rayfield, J.: Dynamic semantic publishing. In: Maass, W., Kowatsch, T. (eds.) Semantic Technologies in Content Management Systems, pp. 49–64. Springer, Heidelberg (2012)
14. Rissen, P.: A Curriculum Ontology [blog post] (2011). http://www.r4isstatic.com/179. Accessed 03 July 2014
15. Waters, J.K.: Sifting the data. T.H.E. J. **40**, 15–18 (2013)

Towards a Linked and Reusable Conceptual Layer Around Higher Education Programs

Fouad Zablith[✉]

Olayan School of Business, Riad El Solh, American University of Beirut,
PO Box 11-0236, Beirut 1107 2020, Lebanon
`fouad.zablith@aub.edu.lb`

Abstract. From a knowledge representation perspective, higher education programs can be exhibited as a set of concepts exchanged in learning
environments to achieve specific learning objectives. Such concepts are
often grouped in different blocks such as courses, modules or topics that
count towards a degree. Currently, most of program representations are
performed at a high level in the form of course descriptions that are part
of course catalogs, or at more specific levels using for example course
syllabi. While this is great for informative purposes, systematically capturing, processing and analyzing the concepts covered in a program is not
possible with this text-based representation. We present in this chapter
a data model based on the linked data principles to create a conceptual
layer around higher education programs. We follow a collaborative approach using a semantic Mediawiki to build the knowledge graph around
a business school curriculum. The impact of this linked open layer is
highlighted at the level of (1) enriching online learning environments by
extending the graph through learning material and selectively pushing
it to existing course pages; and (2) enabling a more in-depth analysis of
the program during review activities.

1 Introduction

Higher eduction is undergoing major changes in the way students and teachers
interact. Such changes are highly dictated by the latest advancement of mobile,
Internet and Web technologies. Those advancements are changing the perception
of information consumers. The expectations from what can be done today in a
classroom, offline as well as online, are increasing. The ease of access to information,
coupled with the ever-improving sophistication of computing devices are opening
new opportunities for educational environments. For example we have seen lately
the increase in the number of open online courses, where teachers can reach out to
thousands of students in a course deliverable. Another example is the availability
of blended learning options that universities are also experimenting with.

While technology is infiltrating higher educational settings, one major question that is yet to be answered is how can existing forms of curricula representation cope with information that is increasingly being exchanged in online
environments? Most higher education curricula are currently represented in

© Springer International Publishing Switzerland 2016
D. Mouromtsev and M. d'Aquin (Eds.): Open Data for Education, LNCS 9500, pp. 86–102, 2016.
DOI: 10.1007/978-3-319-30493-9_5

textual formats, where concepts covered are buried deep into course catalogs and syllabi. This form of representation works well for describing courses and degrees, however it fails when more sophisticated and systematic processing is required. Furthermore, inferring cross connection among courses, or between courses and learning material is a labor-intensive process. This will result in creating boundaries around courses that are hard to break not only in traditional classroom settings, but also in online environments. In other words, a textual representation of curricula cannot keep up and move at the pace of how learning environments are evolving.

We believe that the goodness provided by the latest efforts of the linked data community can be exploited to improve curricula representation. We present in this chapter[1] our effort towards creating a linked layer around a higher education program. We follow the linked data principles [4] to create unique reference-able entities of the concepts exchanged in courses. We propose a data model that we used as a starting point to build the linked data graph. We deploy a semantic Mediawiki [8] to collaboratively generate and inter-connect the linked data graph around a Business School curriculum. We consume the generated data through two pilot applications. The first is in the context of enriching online learning environments, in which we use moodle to showcase how relevant material and information can be easily integrated in course pages without any modifications to the existing platform. The second is by processing the data to present new unprecedented views of the curriculum that can potentially support the program review process at the business school.

This chapter is organized as follows. We first present an overview of existing work in the field in Sect. 2. The data model is then presented and discussed in Sect. 3. Then in Sect. 4 we present the process of generating the linked data graph through the semantic Mediawiki that we implemented. After representing the curriculum through the linked data layer, we present two use-cases of the data in Sect. 5, and conclude in Sect. 6 with potential future research directions.

2   Related Work

The web has been witnessing tremendous changes recently. When the term Semantic Web was first coined by Tim Berners-Lee, it was made clear that the move from links between documents to links between objects will "unleash a revolution of new possibilities" [1]. The objective is to have a web of connections that computers can "understand". In other words entities will be represented and connected through explicit meanings using well-defined vocabularies. Now after around fifteen years, the impact of having "semantics" added to the web is obvious. We have seen governments creating and opening up their data on the web with explicit semantics. For example the "data.gov" platform of the US government that today includes around 164,112 accessible and processable

---

[1] This chapter is an extended version of the paper published in the WWW2015 companion proceedings [11].

datasets. Similarly the UK government initiated their "data.gov.uk" platform for similar purposes, and other governments are now following this trend.

Part of the Semantic Web community effort was the creation of linked data publishing principles [4] that can be summarized as follows:

- To use unique resource identifiers (URI) to uniquely identify abstract concepts and real world objects
- To generate URIs that can be dereferenced on the web through the HTTP protocol
- To publish the data using the Resource Description Framework (RDF) model [6], which is a graph-based data representation that can be easily queried and retrieved
- To have hyperlinks in the form RDF links between entities represented as linked data

The simplicity of such principles contributed to the continuous expansion of the available linked open data cloud [9]. In addition to the government examples mentioned earlier, linked data is getting increasingly adopted in various contexts. Higher education is no exception. In their survey, Dietze et al. [10] highlight the growing adoption of linked data by various universities. They identify the challenges that lie ahead of the technology enhanced learning community for a wider embrace of linked data. This includes (1) the integration of various heterogeneous repositories; (2) the adaptation to the changes underneath services provided such as wed APIs; (3) the mediation and mapping of meta data across educational resources; and (4) interlinking and enriching unstructured data coming for example from text documents. While various research efforts are being invested in the above challenges, educational institutions cannot but benefit from the added value of this linked data layer. In the educational field, various platforms have emerged where linked data is made available for direct consumption and reuse. This includes for example the OU's linked open data platform (http://data.open. ac.uk), the University of Muenster (http://data.uni-muenster.de), the University of Oxford (http://data.ox.ac.uk), the University of Southampton (http:// data.southampton.ac.uk), among others.

Methodologies and frameworks have been proposed to transform existing data sources into linked data [3,12,13]. In this context, available organizational data was transformed, using pre-programmed transformation patterns, into linked data. A LUCERO framework was proposed to perform the following main tasks [13]: *collect* data from various sources of organizational data, where a scheduler automatically checks for data updates through for example Really Simple Syndication (RSS) feeds; *extract* the data and transform it into linked data following pre-defined URI creation rules; *link* the generated data with internal and possibly external data; *store* the data in a triple store and *expose* it for consumption through a SPARQL endpoint.

Following the trend and success of social and knowledge graphs functionalities provided by Facebook[2] and Google[3], there are discussions around the

---

[2] http://newsroom.fb.com/News/562/Introducing-Graph-Search-Beta.

[3] http://www.google.com/insidesearch/features/search/knowledge.html.

value of having an education graph[4] [5]. Heath et al. [5] proposed in their work to create an education graph by processing courses information and learning material from various universities in the UK. In their approach, they mainly rely on bibliographical data of material repositories to identify links to course resources [5].

In our context, we focus in this chapter on interlinking courses within the same institution at the level of concepts covered in course topics. To our knowledge, there is no existing vocabulary that covers this granular information about courses information. To achieve our objective, we had to go at lower (i.e. more specific) conceptual levels by enabling users to have a direct impact on reshaping how courses interconnect among them and with learning resources. This controlled environment is necessary for aligning the conceptual coverage of courses delivered by more than one instructor. We also aimed to have direct input from students to connect to and expand the graph around the education program. For example when students find an interesting material online (e.g. video or article) relevant to a specific course, we wanted to enable them to connect it back to the course by extending the graph and creating the appropriate links to the course.

3   Linked Data Graph

We present in this section our proposed data model to capture course related information. We visualize the data model in Fig. 1. The sources of information can be grouped in two main parts: the first is coming mainly from the courses syllabi, aggregated in the upper part (A) of Fig. 1, while the lower part (B) of Fig. 1 captures information mostly from learning material.

We used the information included in the courses' syllabi template as a starting point to identify the course elements to represent. A typical course syllabus follows a predefined template, which includes the course details such as course number, description, prerequisites, textbook, topical coverage and others. Furthermore, each course has well-defined learning objectives.

We focus in this part on the description of specific aspects that guided the development of the elements that are not covered in the syllabi content. Based on the need for conceptually connecting courses (i.e. beyond the analysis of topics in common), we captured in the graph concepts that are taught at the level of every course topic. This design offers many advantages. First we are getting a more granular view of what is covered in each course. Second, such concepts can be used as anchors between learning material and courses. This will enable course designers to know where exactly each piece of material fits in the course, and this enables learning material to float around not only course topics, but also around the program as a whole. In other words an interesting article that is used in one course, can also be relevant in other courses if the concepts are shared between the courses.

---

[4] http://hackeducation.com/2011/11/10/is-there-an-education-graph.

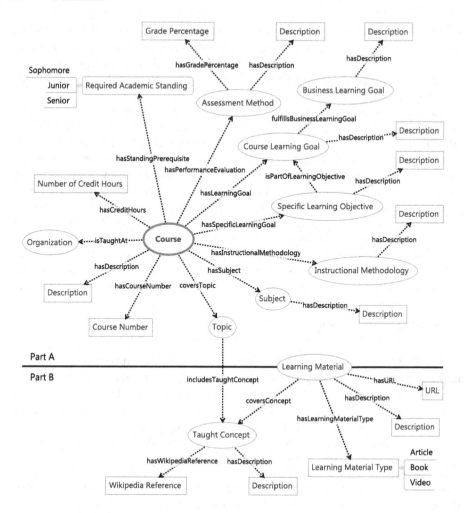

Fig. 1. The linked data graph.

Following the linked data principles [4], we aimed to reuse the available ontologies that are relevant to our context. For example the CourseWare ontology[5] was reused to represent information related to course types, student interaction types, number of credits, assessment methods, and others. The AIISO[6] was adopted to capture the courses' unique codes. We have also used the Dublin Core Terms[7] to represent generic properties such as descriptions. In cases where no vocabularies were found, we created a local vocabulary used in our context.

---

[5] http://courseware.rkbexplorer.com.

[6] http://vocab.org/aiiso/schema.

[7] http://purl.org/dc/terms.

4   Linked Data Generation

After defining our linked data model, we discuss in this part our effort on creating linked data around the higher education program of the School of Business of the American University of Beirut. We followed a collaborative process to build the linked data graph around the program. This involved professors, students and program coordinators. As courses are delivered by different professors, we required to have a system that supports collaboration, through which each professor or involved person can see, expand and modify content as seen appropriate. Another requirement was to enable students to also enrich and add to the graph in a quick and easy way online material that they find interesting. In addition to the above, another needed feature was the ability to control linked data vocabularies. We discuss next our semantic Mediawiki implementation, followed by the steps we took to generate the data using the wiki and a semantic bookmarklet for linking online resources to the courses.

4.1   Semantic Mediawiki Deployment

We have implemented a semantic Mediawiki available at http://linked.aub.edu.lb/collab. For controlling the data generated and vocabularies used, we created several forms that are automatically loaded when new content is to be added. In the case of creating new courses for example, the course form[8] is automatically loaded[9]. The advantage of using such forms is that the user is guided around what fields to fill, and the vocabularies used can be predefined in the form when the RDF is generated.

Linking data is done by controlling the field content in the wiki. By adopting the wiki forms, the fields can guide users in reusing existing concepts from the wiki. For example when specifying the course prerequisite relation, the user is prompted with the list of courses available in the knowledge space that can be chosen from. This is a core feature used for the interlinking process. Another example is at the level of concepts covered in course topics. The same concept can be covered in one or more course topics, creating the links needed.

4.2   Steps Followed for Building the Linked Data Graph

Building the linked data graph was done in three phases. In the first phase, course syllabi are used as entry points, where the high level course information is entered. In the second phase, textbook materials used in the course are processed by the teaching assistants (TAs) to identify the concepts covered in the course. In the third phase, new external materials are added to the graph by instructors and students, using a semantic bookmarklet.

---

[8] http://linked.aub.edu.lb/collab/index.php/Special:FormEdit/Course/New_Course.

[9] For a filled form example, check the "Foundations of Information Systems" course at the following link: http://linked.aub.edu.lb/collab/index.php?title=INFO200_-_Foundations_of_Information_Systems&action=formedit

Phase 1: Creating Courses Information. This phase was straight forward, as existing course syllabi follow a predefined structure. This part was mainly extending Part A of Fig. 1.

Some difficulties were faced at the level of identifying learning goals, as some of the courses learning goals were described as text, without a clear structure. Another complexity at this level was that course learning goals were at two levels (as depicted in Fig. 1). Courses have specific learning goals, which are linked to a broader list of business learning goals identified within our School of Business. This required a two-step data entry to capture linkages among learning objectives.

Another challenge we faced is when the topics of courses change with time, due for example to changes in textbook editions, or in the delivery of content across semesters. For instance, in the Foundations of Information Systems course, the Social Media topic covered in the "Experiencing MIS" [7] textbook has changed from the 3rd to 4th edition. In this case, while the topical coverage has changed, some of the concepts that were covered in the previous edition were still there. We handled such cases by archiving topics and removing the corresponding links to the course. This way we were able to preserve the concepts related to the old topic, and reuse them if needed in the new topic.

The evolution of changes at this level can be better managed in the future. One possible improvement can be done through capturing temporal changes, coupled with the type of changes performed (e.g. adding or removing concepts from the concept graph). While tracing such evolution patterns can be done at the data entry level, a post analysis of changes occurring on the data graph can be possibly performed. Such features are beneficial for analytics applications, and can be further explored as part of our future research. We have processed so far the 19 core courses offered at the School of Business, leaving the elective courses to be represented at a later stage.

Phase 2: Identifying Concepts Taught in Courses. This phase was the most extensive and time consuming phase. The focus at this level was on capturing the concepts covered in the core learning material of courses such as textbooks mentioned in the courses syllabi.

TAs were trained to identify concepts covered in the topics of the courses. Based on the model we created (cf. Part B of Fig. 1), concepts are linked to the topics of a course, and not directly to the course itself. This choice of design enables grouping concepts by topic, rather than by course. This has a practical implication in filtering concepts covered in specific topics (as we discuss later in the chapter when we integrate data in moodle). Another implication is at the level of program analytics. Overlap among course topics can be highlighted easily, which proved to be useful for the program review and design exercises.

At this level, the TAs were going into each topic covered in the textbook to identify the main concepts, adding definitions from the book, and linking to a Wikipedia[10] reference when found. When applicable, the TAs were instructed

---

[10] www.wikipedia.org.

to reuse existing concepts in the repository. The *concepts* field in the *topic* form automatically provides the TAs with the list of existing concepts to choose from.

One challenge at this level is when concepts from different topics share the same name, however are semantically different. For example, the *Optimization* concept covered in the *Managerial Decision Making* (i.e. operations research), is semantically different from the optimization concept in the *Managerial Economics* field. This is where Wikipedia is used as an external reference for disambiguating such cases. While linking the concepts to Wikipedia entries is currently done manually, this task can potentially be performed (semi)-automatically by querying DBpedia [2] and aiming to find overlap between the concepts' definition and the description in the DBpedia page. In addition to text matching, the graph of the concepts can be used as a context to highlight the degree of matching. Then the user can browse the proposed matching concepts to select the most appropriate one.

Currently concepts identified are not related through an explicit relation. We are planning to capture in the future relations such as sub-class and prerequisite relations among concepts. This can also have an impact on the analysis of course sequencing in the program. So far we have identified around 2,680 concepts[11] covered in the core courses of the School of Business.

Phase 3: Semantically Anchoring Learning Material to Courses. While the previous phases were mainly focusing on reorganizing internal knowledge sources, this phase is more about linking new materials to the program, by anchoring them to concepts covered in courses. Here the aim is to enable students and instructors to link interesting online material to the graph.

We developed a simple bookmarklet that can be installed in any browser. When the user links a learning material, clicking the bookmarklet will automatically extract the page link, title and description. The user is then prompted to a page pointing to our wiki platform, where the concepts covered in the material can be entered by reusing existing concepts from the graph[12].

As mentioned earlier, links between a learning material and the program are done through the concepts. This somehow enables educators and students to think more around the relevance of the material around concepts. For example we witnessed a student who bookmarked an article related to Big Data[13], relevant to the Foundations of Information Systems course she was enrolled in. While anchoring this article, she reused concepts from the graph that indirectly spread to two other courses. Starting from an information systems' related material, the student indirectly linked to a management and operations management courses. The full list of concepts highlighted by the student can be found in Fig. 2.

---

[11] The full list of concepts can be accessed at: http://linked.aub.edu.lb/collab/index.php/Category:Learning_concepts.

[12] A video tutorial on how to use the bookmarklet can he accessed at: http://linked.aub.edu.lb/docs/tutorial_material_bookmark.

[13] http://www.capgemini.com/resources/the-deciding-factor-big-data-decision-making.

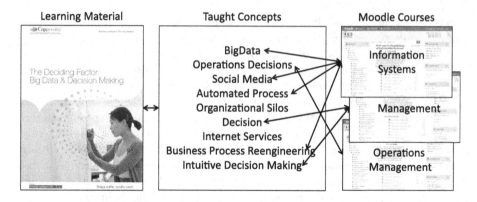

Fig. 2. Learning material cross connecting courses.

5   Consumption of Linked Data

We present in this section two scenarios where we used the linked data generated. We accessed the data through the semantic search feature of the wiki endpoint that provides querying functionalities with different output formats. The queries were formulated and passed through php to the wiki query endpoint, and results were returned in JSON for processing.

5.1   Interlinking Course Learning Material on Moodle

One challenge that web developers face when customizing existing platforms (e.g. moodle), is the need for understanding the existing code infrastructure to be able to extend the system's functionalities. Having linked data provides several advantages, including data portability and reuse. The ease of data extraction and consumption through endpoints such as SPARQL endpoints or the wiki's semantic search features can somehow break such development barriers. In our context we show that, when the data is decoupled from the application layer, it was possible to enrich moodle pages without having to modify the corresponding source code.

The linked data generated was used to enrich and connect the moodle course pages. Moodle is extensively used at the School of Business as a way to communicate course related information, and to interact around course deliverables. The suggested design of moodle course pages at the American University of Beirut is to subdivide the page using the covered topics in the course, and add topic related material, assignments and other activities under each section. Course instructors design their own page at the beginning of each semester, or reuse an existing one if the course was already taught by the instructor. One trend that is observed is that instructors of the same course tend to share interesting materials that could be used in classrooms. However such insights are not usually captured, and have to be re-shared whenever a new instructor teaches the course. In addition to sharing constraints, another bottleneck observed is that courses are designed (even on moodle) in isolation. However material relevant to some courses, can potentially be relevant to others (as perceived with the Big Data article discussed earlier).

The aim of this application is to break out of the static and isolated nature of content shared within a course topic on moodle. The application offers the functionality to dynamically enrich moodle pages, without leaving it, with material relevant to the topics of the course. Figure 3 highlights the steps performed within the moodle page.

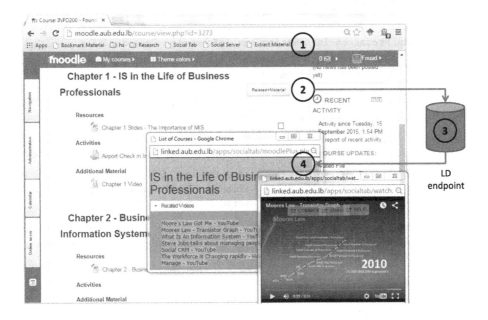

Fig. 3. Dynamically enriching moodle course pages.

A video tutorial is available as a guide for students and instructors to follow[14]. This application can be launched by pressing a bookmarklet in the browser (Part 1 in Fig. 3), and the following sequence of steps is performed:

1. Scan the topics available in the moodle page: the executed code first will launch a javascript to scan the moodle page for the topics header. This code snippet extracts the section names by filtering the HTML class names that match the ones provided by the moodle page.
2. Enable the buttons on the moodle page: the javascript will dynamically inject form buttons next to each course topic on the moodle page (Part 2 in Fig. 3), with the corresponding course code and topic embedded inside the button links.

---

[14] http://linked.aub.edu.lb/docs/tutorial_extract_material.

3. Send query to the wiki linked data endpoint: when the user presses the button, a query is passed to the wiki endpoint (Part 3 in Fig. 3) with the topic and course code. As per our model in Fig. 1, the link between the course and material is done at two levels, through the *topic* and then through the *taught concepts*. Hence the query is built to first fetch the concepts covered in a topic, and then filter the learning material based on the concepts in focus.

4. Parse and visualize query results: the query results are returned in a JSON output, and parsed to identify the different types of related material (so far we have video material, articles, and books). Below is an example of a JSON output linking the topics to concepts:

```
"results": {
"Business Processes Information and Information Systems": {
    "printouts": {
        "Includes learning concept": [
        {
            "fulltext": "Business Process",
            "fullurl": "http:\\linked.aub.edu.lb\collab\
                index.php\Business_Process"
        },
        {
            "fulltext": "Automated Process",
            "fullurl": "http:\\linked.aub.edu.lb\collab\
                /index.php\Automated_Process"
        }...
```

The following shows the JSON output from the second query extracting the learning material and their concepts:

```
"results": {
"Amazing mind reader reveals his 'gift' - YouTube": {
    "printouts": {
        "Covers concept": [
        {
            "fulltext": "Social Media",
            "fullurl": "http:\\linked.aub.edu.lb\collab\
                index.php\Social_Media"
        },
        {
            "fulltext": "Social CRM",
            "fullurl": "http:\\linked.aub.edu.lb\collab\
                index.php\Social_CRM"
        },
        {
            "fulltext": "Privacy",
            "fullurl": "http:\\linked.aub.edu.lb\collab\
                index.php\Privacy"
        }
        ]
    },
```

```
"fulltext": "Amazing mind reader reveals his 'gift' - YouTube",
"fullurl": "http:\\linked.aub.edu.lb\collab\index.php\
   /Amazing_mind_reader_reveals_his_
  }...
```

Finally the results are used to populate the page where users can read articles or play videos (Part 4 in Fig. 3).

Following a concept centric design in our proposed model, coupled with the portability and ease of reuse of linked data, enabled us to implement a solution to dynamically enrich learning environments with relevant material. This design can automatically place learning material under the right topic on the course pages. For example going back to the Big Data article mentioned earlier in Fig. 2, this article will dynamically appear in the information systems, management and operations management moodle pages.

5.2   Using Linked Data for Program Review

The second scenario in which we relied on the linked data generated from this work is in the program review process at the School of Business. Every four years, the curriculum has to be reviewed for changes, where courses are studied to be added or removed from the program. Another task that is part of the review process is the course sequencing. To achieve this purpose, traditionally each course is studied on its own, and compared to other courses, and to the learning objectives of the program. However it was not possible to perform an in-depth analysis beyond the syllabi content, and hence it was hard to know exactly what is covered in each course, and what are the concepts that are repeated across different courses.

The ability to have an overview of what is covered in a program and in which course is important for the program review activity. Hence course mapping was highlighted to be one of the major tasks required for adjusting course sequencing and coverage. Currently course sequencing is done at very high level based on the topics covered. However over the years, with some changes that occur in the content and course delivery, such sequencing should be revisited regularly. Our platform provided an unprecedented view of how courses overlap, down to the concept level.

We implemented a visualization showing how courses connect through topics and taught concepts. This visualization is dynamically generated based on the wiki content. A visualization example around the "Foundations of Informations Systems" course is presented in Fig. 4.

It highlights how the "Database Marketing" topic covers the "Decision Tree" concept, which is also repeated in the "Game Theory and Strategic Behavior" topic of the "Managerial Economics" course. This visualization can be accessed online at: http://linked.aub.edu.lb/collab/index.php/ Learning_Concepts_Graph, and is created using the Semantic Graph extension[15],

---

[15] http://semanticgraph.sourceforge.net.

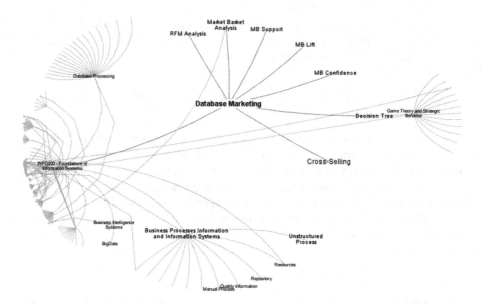

Fig. 4. Concept map around the foundations of information systems course with a focus on the "Database Marketing" topic.

with the HyperGraph layout[16]. With the presence of semantic relations, this visualization was easy to implement. The extension requires the following to be specified: the wiki resource, the semantic relations to extract, the depth of relations and the layout dimensions. For example to generate this graph, we set the course page in focus, and used the "covers topic" relation to extract the related topics, and the concepts were selected based on the "includes learning concept" relation. Informal feedback from faculty members highlighted the complexity of this view in analyzing the concepts covered in a systematic way. A simpler representation was required to easily identify the repetition of concepts across courses.

A tabular view was identified as a potential representation of courses, their topics, and corresponding concepts. We implemented a table on top of two queries. One fetches the course to topic relationships, and another extracts the topic to concept relations. We stored the results in JSON format, and loop through the entities to detect overlaps. We built a simple HTML-based table to render the results. The table can be accessed online[17]. We show part of the table in Fig. 5. The users were able to see the list of courses (course codes are on the top row of the table), and in each row we present the list of topics and their corresponding concepts. The "X" marks the occurrence of this concept in the whole program. Due to the large size of the table, and when more than one "X" is in the cell, the user can roll the mouse over the table cell to see where this

---

[16] http://hypergraph.sourceforge.net.
[17] http://linked.aub.edu.lb/apps/tablebrowser/table.php.

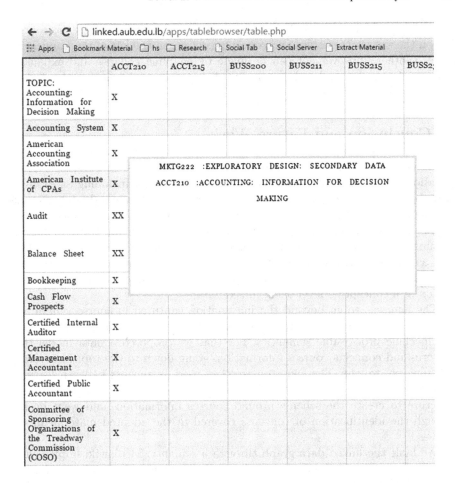

Fig. 5. Tabular representation of concepts covered in courses with overlap detection.

concept occurs in other courses. For example Fig. 5 indicates that the concept *Audit* is mentioned in the Accounting 210, and in the Marketing 222 courses. We also fetch the topics where the concepts are covered.

This tool proved to be useful in the program review process, as it enabled the program coordinators and instructors to highlight the parts of the courses that require adjustment. This table implementation can benefit from various improvements. For example we anticipate the need for creating filters for users to specify a specific set of courses to display. For example if one of the departments at the School of Business is interested in how courses in their subjects overlap, a filter can be added based on the subject, and the table will only highlight courses that fall within this subject. Other filters could be based on learning objectives, to highlight how concepts feed into such objectives. In addition to improvements based on the data selection features, we see further improvements that can be made at the interface level. For example with the long list of courses and subjects, some aspects of the table can be made static such as the courses row on the top, or the topics and concepts column on the left of the application page. Another improvement can be made by enabling collapse and expansion features of concepts under the topics. Such features can be easily enhanced by using more sophisticated script-based modules.

6 Conclusion and Future Work

Higher eduction programs are represented in different ways. Traditionally representations were mainly performed through text documents including for example course syllabi and program catalogs. However such text layers do not allow us to exploit and dig deeper in the knowledge exchanged in higher education. As a result courses are mainly designed in isolation, with the aim to fulfill specific learning objectives. Furthermore, semantic connections between course entities such as learning material and topics are hard to systematically infer. We presented in this chapter our effort on connecting higher education program information at a conceptual level using linked data.

The aim was to go beyond the information captured in course syllabi and catalogs. We required to go at a more granular conceptual level with explicit and machine processable semantics. For that we proposed a data model that captures and connects courses information, going down to the topical coverage and concepts taught. The concepts are then used as anchors between learning material and the higher education program. We relied on the existing syllabi structure to create the schema around courses information, and went deeper through the identification of concepts covered in the adopted material in the courses.

We built the linked data graph through a semantic Mediawiki implementation. The aim was to collaboratively expand this knowledge layer by involving various faculty members and students at the School of Business. The wiki offers a platform where instructors can reach a consensus around what is taught in their courses, by having a controlled environment to manage the reuse of existing knowledge and appropriate vocabularies for creating linkages. We created a semantic bookmarklet for learning material through which users can directly bookmark an interesting learning resource and be redirected to the wiki with existing elements automatically extracted from the page.

We also introduced how this interconnected data layer around the curriculum can help in the review and design of higher education curricula. The deep links among courses can be visualized in different formats. We presented two in this work, (1) a concept map that shows the connections around courses, topics and covered concepts, and (2) a table that highlights the occurrence of concepts in the program. This work offered program designers at the School of Business a unique view that was not possible before on how courses conceptually connect. They were able to see how concepts are repeated in courses, enabling them to make better decisions around required changes in the program.

Linked data provides several benefits in the management of curriculum and educational data, we discuss three advantages herein. The first advantage is related to the native graph-based structure of the data model. Such representation enables an easy extension of the model whenever required. This is one of the crucial requirements for keeping up with the dynamic nature of curriculum content. The second benefit lies at the level of portability and ease of data reuse in different contexts as highlighted in our scenarios. When semantics are explicitly encoded in the data, building applications on top of it is made easy. For example we were able to selectively extract parts of the knowledge around courses to visualize course contents in the form of tables or diagrams. We also showed how we can bring learning material in the context of a course, and place them under its corresponding topic within existing online learning environments such as moodle. The use of taught concepts as anchors between the materials and courses gave us a great flexibility in fetching material that stretch the boundaries of courses delivery, which can improve the learning experience of students and highlight how courses cross-connect in the degree program they are enrolled in. The third advantage of the use of linked data is driven by the growing number of the available tools that are supporting the storage, maintenance and extraction of linked data. This trend will engage data publishers in adopting linked data principles as a way to serve their data. Hence in our context, we know that the availability of our conceptual layer will have a longer life span. We also believe that the value of this data and the sophistication of applications built will increase, when the data is combined with external linked data sources.

There are different research directions to follow next. One natural continuation of this work is to evaluate the impact of this linked representation of courses on curriculum changes and reviews. We are currently developing further visualizations, and planning to evaluate the degree of insights that can be generated from such diagrams. We are also interested in evaluating the impact of having such data layer in learning environments through a guided user study coupled with evaluation measures. Another line of work we are currently pursuing is on capturing social interactions around our education program. By merging the social and education graphs, we anticipate that we will be able to granularly analyze how teachers and students interact around concepts delivered during their higher education journeys. This new linked data layer will offer endless opportunities in manipulating curricula related information in new and insightful contexts.

Acknowledgments. This work was supported by the University Research Board grant at the American University of Beirut. I would like to thank my colleagues and students who contributed to the extension of the graph through the wiki, and supported me in the implementation of the tools. I would also like to thank the students who have used the bookmarklet to bookmark the interesting materials that triggered invaluable insights and observations.

References

1. Berners-Lee, T., Hendler, J., Lassila, O.: The semantic web. Sci. Am. **284**(5), 28–37 (2001)
2. Bizer, C., Lehmann, J., Kobilarov, G., Auer, S., Becker, C., Cyganiak, R., Hellmann, S.: DBpedia-A crystallization point for the Web of Data. Web Semant.: Sci., Servi. Agents World Wide Web **7**(3), 154–165 (2009)
3. d'Aquin, M.: Linked Data for Open and Distance Learning. Commonwealth of Learning report (2012)
4. Heath, T., Bizer, C.: Linked data: evolving the web into a global data space. Synth. Lect. Semant. Web: Theor. Technol. **1**, 1–136 (2011)
5. Heath, T., Singer, R., Shabir, N., Clarke, C., Leavesley, J.: Assembling and applying an education graph based on learning resources in universities. Workshop, In: Linked Learning (LILE) (2012)
6. Klyne, G., Carroll, J.J.: Resource description framework (RDF): Concepts and abstract syntax. Technical report (2006)
7. Kroenke, D., Bunker, D., Wilson, D.N.: Experiencing Mis. Prentice Hall, Upper Saddle River (2010)
8. Krötzsch, M., Vrandečić, D., Völkel, M.: Semantic MediaWiki. In: Cruz, I., Decker, S., Allemang, D., Preist, C., Schwabe, D., Mika, P., Uschold, M., Aroyo, L.M. (eds.) ISWC 2006. LNCS, vol. 4273, pp. 935–942. Springer, Heidelberg (2006)
9. Schmachtenberg, M., Bizer, C., Jentzsch, A., Cyganiak, R.: Linking Open Data Cloud Diagram (2014)
10. Dietze, S., Sanchez-Alonso, S., Ebner, H., Yu, H.Q., Giordano, D., Marenzi, I., Nunes, B.P.: Interlinking educational resources and the web of data. Prog. **47**(1), 60–91 (2013)
11. Zablith, F.: Interconnecting and enriching higher education programs using linked data. In: Proceedings of the 24th International Conference on World Wide Web Companion, pp. 711–716. International World Wide Web Conferences Steering Committee (2015)
12. Zablith, F., d'Aquin, M., Brown, S., Green-Hughes, L.: Consuming linked data within a large educational organization. In: Proceedings of the Second International Workshop on Consuming Linked Data (COLD) at International Semantic Web Conference (ISWC) (2011)
13. Zablith, F., Fernandez, M., Rowe, M.: Production and consumption of university linked data. Interact. Learn. Environ. **23**(1), 55–78 (2015)

Collaborative Authoring of OpenCourseWare: The Best Practices and Complex Solution

Darya Tarasowa[1,2(✉)] and Sören Auer[1,2]

[1] University of Bonn, Bonn, Germany
darya.tarasowa@gmail.com, soeren.auer@gmail.com
[2] Fraunhofer IAIS, Bonn, Germany

Abstract. The lack of high-quality educational resources is a key issue on the way to success of the OpenCourseWare movement. However, the state-of-art approaches of producing such content demand a lot of resources, thus limiting the percent of such courses in a total amount of content available. An important step towards decreasing costs while increasing the quality of the educational material is applying collaborative techniques to the production process. Such collaboration affects different aspects of OpenCourseWare production, such as: content annotation, personalization, sharing and other. In the current paper we aim to investigate the state-of-art of the collaborative authoring of Open-CourseWare in all its aspects, finding out the major gaps and the most promising approaches to fulfill them. Based on the study results we developed Slidewiki - an example application for the OpenCourseWare collaborative authoring that implements the most promising solutions for the gaps found.

Keywords: Collaborative authoring · Educational content · Survey · State-of-art · Analysis

1 Introduction

An important kind of educational open data is OpenCourseWare (OCW). This type of data requires the content to be presented as a combination of reusable and remixable learning objects.

Leinonen et al. in their study [29] provide requirements for the learning objects which can satisfy the learning needs:

- Learning objects should be relevant to the learner and thus easily modified to fit the learner's needs.
- They should be of good quality and contain no factual errors.
- They should disclose their point of view and in the case of science be free from bias.
- They should not have hidden costs or prohibiting limitations on use.
- A good learning resource should also be able to "travel well", to be easily translated and re-contextualized.

© Springer International Publishing Switzerland 2016
D. Mouromtsev and M. d'Aquin (Eds.): Open Data for Education, LNCS 9500, pp. 103–131, 2016.
DOI: 10.1007/978-3-319-30493-9_6

Another four factors of OCW development success are listed in [41] and include:

- Convergence toward common metadata
- Balancing expert and community definitions of quality
- Community input
- Interoperability

Due to those requirements, the effective state-of-art approaches of producing such content, for example, Massive Online Open Courses (MOOCs), demand a lot of resources. This limits the percentage of such courses in a total amount of content available. At the same time, the number of low-quality learning objects grows fast, making the content search and filtering a challenging and exhausting task. Thus, the lack of high-quality learning objects is a key issue on the way to success of the OpenCourseWare movement. According to statistics [36], the start-up capital of OpenLearn[1] amounts to £5,650,000 (up $9.9 million), and the cost of upgrading the MOODLE[2] platform is also large. MIT OCW costs approximately $3.5 million every year, which means the release of each course took about 100 working hours in average.

An important step towards decreasing the costs while increasing the quality of the educational material is applying collaborative techniques to the production process. Such collaboration affects different aspects of content production such as content annotation, personalization and sharing. According to the study [42], already at that time most of the work in academia, business and industry was completed by groups of people collaboratively. This is why collaborative creation of education materials is natural. From psychological and sociological point of view it is proven, that people like to collaborate [20,49]. Several psychologists demonstrated the effects of collective work on the process of cognitive development of people. Generally, people even *prefer* to work together. According to [20,49], the reasons for this include:

- benefiting from partner's knowledge;
- building strong personal relations with others;
- increasing the work quality by critics and experience exchange;
- avoiding duplication and redundancy of tasks.

However, if we look at the materials available on the Web, many of them do not satisfy the quality requirements discussed above and therefore do not satisfy learner needs. Considering, that collaboration itself can not be a reason for that, the only reason can be the technological issues, disturbing (groups of) teachers from doing their best.

In order to define and deal with the issues, we conducted a comprehensive study of the state-of-art in collaborative authoring of reusable educational materials. Additionally, we were motivated by the fact, that although many of the approaches found in the literature were proven to be beneficial, the existing leading OCW authoring platforms do not integrate them. We aimed to collect and

---

[1] http://www.open.edu/openlearn.
[2] http://moodle.org.

describe these approaches in order to attract community attention to them. We describe our research method for paper selection and analyze the state of art in Sect. 2. We define major gaps and promising approaches to fulfill them, as well as common terms used in the area in Sect. 3. In order to examine the approaches on practice, we have developed SlideWiki - an experimental web-based platform for the collaborative OCW authoring. We describe, evaluate and discuss the platform in Sect. 4. Finally we conclude our work and propose the directions for future work in Sect. 5.

2 Collaborative OCW Authoring State-of-art Analysis

2.1 Organization of the Study

In order to ensure the completeness of the study, we followed a formal systematic literature review process based on the guidelines proposed in [17,27]. As a part of the review process, we developed a protocol (described in the sequel) that provides a plan for the review in terms of the method to be followed, including the research questions and the data to be extracted.

Research Questions. In order to organize the survey we formulated four main research questions:

1. *What are the main challenging tasks in collaborative authoring of reusable OCW?*
2. *Which technologies are being used to solve these tasks?*
3. *How well covered are the challenging tasks in the scientific literature?*
4. *What are the main gaps in state-of-art research?*

Search Strategy. To cover as many relevant publications as possible, we used the following electronic libraries:

- ACM Digital Library
- IEEE Xplore Digital Library
- ScienceDirect
- Springerlink
- ISI Web of Sciences

Based on the research questions and pilot studies, we found the following basic terms to be most appropriate for the systematic review:

1. *crowd-sourcing* OR *crowdsourcing* OR *collaboration*
2. *collaborative* OR *collective*
3. *authoring* OR *creation* OR *edit* OR *editing*
4. *education* OR *educational* OR *learning* OR *e-learning*
5. *content* OR *resources* OR *material*

To construct the search string, all these search terms were combined using Boolean "AND" as follows:

$$(1 \text{ OR } (2 \text{ AND } 3)) \text{ AND } 4 \text{ AND } 5$$

The next decision was to find the suitable field (i.e. title, abstract and full-text) to apply the search string on. In our experience, searching in the 'title' alone does not always provide us with all relevant publications. Thus, 'abstract' or 'full-text' of publications should potentially be included. On the other hand, since the search on the full-text of studies results in many irrelevant publications, we chose to apply the search query additionally on the 'abstract' of the studies. This means a study is selected as a candidate study if its title or abstract contains the keywords defined in the search string. In addition, we limited our search to the publications that are written in English and are published after 1999, when, according to [16], "the first glimmerings of Web 2.0 were beginning to appear".

Study Selection. Based on the search query discussed above, we have collected 4904 papers. We have imported the lists of papers received from the libraries into MicrosoftExcel in order to proceed filtering.

The filtering was proceeded as follows:

1. Remove duplicates by ordering the titles alphabetically and using in-built Excel functions.
2. Filter the titles, keeping only those which imply the conformity of the paper content to the research questions. As this might result in the exclusion of important papers, we additionally took step 5 to ensure the presence of all significant papers in the field.
3. Abstract filtering.
4. Full-text filtering.
5. Enriching from the references in order to ensure the presence of all significant papers in the field. The references to be included were selected based on the rules for filtering the titles from the initial scope. Before including the referenced paper, it was ensured that it was not already present in the scope.
6. Filtering "low-level" papers.

2.2   Overview of Included Studies

Within the 131 selected papers there were in total 23 surveys, experiment studies and essays related to our topic of interest. A significant part of the surveys found is focused only on one particular aspect or technology (e.g. "Social Networking" or "Collaborative adaptation authoring"). We incorporated the most important findings of the surveys into the related sections. The noticeable and influential essays and experiment studies results are discussed/cited mainly in Sects. 1 and 3.

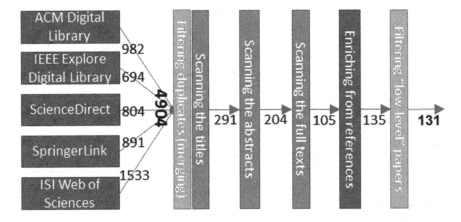

Fig. 1. Steps followed to scope the search results.

Additionally we found several surveys aimed to answer research questions similar to the ones we defined. However, the here presented analysis of the studies shows the absence of detailed and systematic surveys, and thereby proves the actuality of our research.

Thus, the study the existing authoring tools [18] in the related work section research the field considering five dimensions: (1) open hypermedia compliant systems, (2) design metaphors used to create the systems, (3) semantic characteristics, (4) collaborative characteristics and (5) adaptive and intelligence characteristics. The nature of the work however does not allow the authors to present a comprehensive analysis of existing systems (it is not a survey, but system description paper). Neither does the survey include the latest findings in the field due to its date (Fig. 1).

Another review [5] studies the collaborative systems in the understanding of that time. This essential study provides detailed functional overview of the systems supporting collaborative work. The authors define main classes of collaborative systems and classify the studied applications accordingly. The survey however is not focused on educational materials authoring and is out-dated.

Several surveys [6,50] discuss pedagogical and sociological aspects of e-collaboration rather than technical. Thus, in his paper [6] the author provides the classification of e-collaboration types, (e.g. synchronous versus asynchronous collaboration, continuous versus one-time contribution approaches etc.). Due to the nature of the survey, significant attention is paid to user motivation. The paper is important to understand social behavior of contributors, but addresses the e-collaboration from a different aspect than the current study.

The study [33] gives an overview of state-of-art in the collaborative authoring of OCW. Although the research questions are similar with those of the current study, the approach is different. The researchers interviewed the end users of the OERs (teachers) and summarized their experiences and the challenges they met. According to the approach, the study can not be considered systematic.

Table 1. Surveys and essays studying technologies in application to collaborative OCW authoring

Concept	Surveys
Authoring paradigms	
crowdsourcing	[35,41]
wiki	
workflow-based	
Semantic Web technologies	
semantic wiki	[37,39]
ontologies	[10]
RSS-feed	
Social networking	
activity feed	
communication tools	[5,37,46]
social annotation	[19,28,43]
social voting	
social games	
Other	
AI	
tree-structure	
grids, LORs, mash-ups	[55]
Total	**12**

Table 2. Surveys and essays addressing the issues of collaborative OCW authoring

Concept	Surveys
Co-creation	
Content authoring	
Metadata authoring	[38]
Assessment items authoring	
Quality assurance	
Categorization	
Personalization	[8,10,28]
Localization	
Reuse and re-purpose	
Search, aggregation, filtering	
Remixing	
Organization	
Social Collaboration	
Negotiation	[26]
Awareness	[26]
Network building	[6,26]
Engagement	[6,26]
Total	**6**

It is nevertheless important from the practical point of view. The recommendations given by the authors in the conclusive part however lack technical depth and are too general to be directly incorporated by researches and developers on a technical level.

Other surveys and essays that address specific aspects or technologies of collaborative OCW authoring are summarized in Tables 1 and 2.

2.3   Study Results and Conclusions

The first step of our analysis was the identification of aspects of collaborative OCW authoring. We then classify the papers according to the aspect representing the main focus of the respective paper. We refine each aspect with additional sub-branches, thus creating a mind-map. For aspects without associated papers, we tried to identify additional papers through references from the included papers. We also collected information about the used technologies in the mind-map. Subsequently we created a matrix indicating the papers with regard to the aspects and technologies they address. Finally, the matrix helps us to identify gaps and promising areas of further research. We now describe these steps in more detail.

OCW Collaborative Authoring Mind Map. Our initial mind map included root branches for 11 concepts, some of which were further branched. However, as the detailed study of the selected papers has led to significant changes, we do not illustrate the initial map here, limiting ourselves to a brief discussion of findings made already at this initial step[3]. The mind map showed the main research trends in the field, as well as underrepresentation of some aspects, which we assume to be relevant and important to the field. The distribution of the papers between the concepts is presented at Fig. 2. There is an important assumption here, that each paper is related to just one concept (which is not true in the majority of cases). This assumption is only initially made at this point for simplifying the usage of the map.

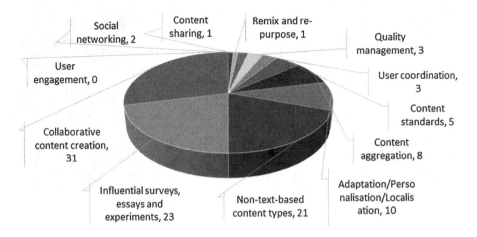

Fig. 2. Distribution of selected papers between a priori identified concepts

As can be seen on the chart, more than one fifth of the papers selected are surveys essays or case studies (23 articles). We have separated these papers from the rest as they can not be assigned to any aspect. As well, we have separated research related to non-text-based content, such as video/audio records, graphics or three-dimensional models (21 articles). This is due to the significant difference in approach used to deal with these types of content in comparison with text-based formats.

After separating these two kinds of papers, we observe that the majority of the remaining papers focus on the aspect of content development using *crowdsourcing* or *wiki* paradigms (31 articles). The aspects of content adaptation/personalization and content aggregation receive decent attention as well (10 and 8 papers respectively). Less than 10 percent of the selected papers are focused on any of the other aspects and we could not identify a single paper making its main contribution in the user engagement aspect of the OCW collaborative authoring.

---

[3] The final map is discussed further and presented at Fig. 3.

Table 3. OCW collaborative authoring papers distribution. The numbers in the first column indicate technologies used to solve issues in the rows: (1) crowdsourcing, (2) wiki, (3) workflow-based authoring, (4) semantic wiki, (5) ontologies, (6) RSS-feed, (7) activity feed, (8) communication tools, (9) social annotation, (10) social voting, (11) social games, (12) AI, (13) tree-structure, (14) LORs, grids, mash-ups

	Content co-creation							Reuse and re-purpose			Social collaboration				
	Content authoring	Metadata authoring	Assessment items auth.	Quality assurance	Categor-ization	Personal-ization	Local-ization	Search etc.	Remixing	Organ-ization	Negotiation	Awareness	Network building	Engagement	Total
Authoring paradigms															
1	6	3	3				3	2							16
2	16						1		1	2	16				16
3	6			1											6
Semantic Web technologies															
4	5	1				1		5			5				5
5				2				1		4					6
6								6	1						7
Social networking															
7		1					3	2				3	1		9
8				2							22	6	1	1	15
9		17	2	5	6	3	1	17			1		2	1	20
10				2		1							1	1	5
11														1	1
Other															
12	3	1		6	4	4		1			1				15
13										5					5
14								3	3						3
To-tal	30	18	5	13	9	10	5	26	4	10	28	10	3	3	65

OCW Collaborative Authoring Matrix. During the detailed analysis of the selected papers we created a matrix indicating the distribution of papers between aspects versus employed technologies. Again, we do not consider the papers focusing on non-text-based content, as well as surveys or essays. The matrix shows the (proposed) use of technologies with regard to the OCW authoring aspects discussed in the selected papers. Each cell in the matrix includes the papers in which the technology (indicated by the row) was applied to the aspect (indicated by the column). As articles usually cover multiple aspects and technologies, they might occur multiple times in the matrix. Due to the size of the matrix, we present a simplified version in Table 3. Each cell here indicates the *number* of papers (instead of paper identifiers list) addressing or proposing the application of the corresponding technology to a corresponding task. The total number of papers in the black row and column considers every paper only once.

The filling out of the matrix allowed us to build the final version of OCW mind map, presented in Fig. 3. The mind map represents the main concepts and technologies of the field. Together with Sect. 3 it can be used for a speedy entering of the field for beginning researchers.

OCW Collaborative Authoring State-of-art Gaps. The OCW collaborative authoring matrix gives additional details on the most and least researched

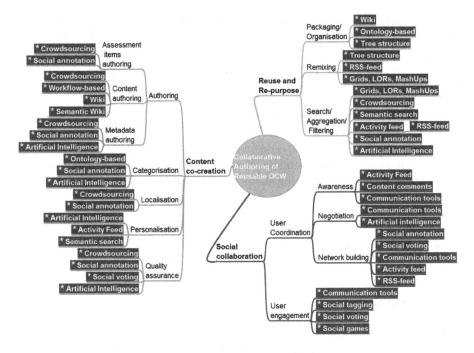

Fig. 3. The mind map for Collaborative authoring of reusable OCW

topics in the area. For example, while 30 of the studied practical papers studied discuss different approaches to collaborative content authoring, only 13 of them address the quality assurance issue. In this subsection we discuss the main gaps of the state-of-art research.

Content localization issue. Although it is believed that OpenCourseWare brings the most benefits to the developing countries, the translation and localization approaches are almost ignored in the published research. This challenge is well-known and often discussed but still no satisfactory solution has been provided. The power of crowdsourcing could be used to translate the content with reasonable quality, but this raises the problem of content synchronization and cultural barriers.

Content remixing. The issue of content remixing is also often ignored in the state-of-art research. The educational content has to be re-designed from year to year, often new topics should be included, and parts of different courses might need to be combined. Although several approaches exist to address this challenge using mashups and LORs, these solutions do not involve content versionning and crowdsourcing. An interesting approach is proposed in [4]. It uses RSS-feed and machine-learning algorithms for filtering the content from the feed and assembling it according to the user preferences. A mashup created in such way can be further edited by the collaborators.

Network building and User engagement. These two aspects are interconnected and are important for the success of any collaborative system. Although several strategies are available from a technological side (like digital badges or social gaming), only a few approaches implement and evaluate them. The study of these aspects lack especially evaluation and comparison of the existing methods.

Complex approaches. Additionally we have noticed the lack of complex approaches to the collaborative OCW authoring. We claim that the defined aspects are highly interrelated and every approach addressing one of these aspects must be studied for its influence on the others. For example, the approach of collaborative content authoring should not be proposed without a solution for the content quality issue, which is impossible to provide without proposing a user coordination scheme. Without an effective publishing approach it is impossible to make the developed learning objects truly reusable. That is why a complex solution has to provide tools for preferably collaborative metadata authoring as well.

3   Terminology

During our study we faced plenty of ambiguous terms and domain-based concepts that might be unclear for non-education specialists, for example, computer scientists aiming to implement an approach. In this section we focus on such terms, collecting their definitions and making our own conclusions on their appropriate usage.

3.1   Learning

E-learning - a learning conducted on the Internet [13]. A wide set of applications and processes, which use available electronic media and tools to deliver vocational education and training [14].

SMLearning - type of e-learning that assumes the functions of a Social Media platform and extends its features to educational context [11].

Blended Learning - the (organic) integration of online digital learning and face to face classroom learning [9].

Adaptive Learning - learning that enables learners to customize their learning environments and dynamically adapts learning content to learners' learning needs [7]

Virtual Attendance - the combination of synchronous and asynchronous ICT tools used to provide distance-education students with the same educational experience that conventional students receive in facetoface (henceforth, F2F) taught classes.

3.2 Content

Learning Object - a small, reusable digital component that can be selectively applied - alone or in combination - by computer software, learning facilitators or learners themselves, to meet individual needs for learning or performance support [48]. A later IEEE definition states: any entity digital or non-digital, which can be used, re-used, or referenced during technology-supported learning [32]. In the context of Semantic Wikis [31] expands the term to include any real world objects such as people, places, organizations and events.

Learning Object Repository (LOR) - a general term for an online collection of learning objects.

OCW - a combination of learning objects, organized in a structured way according to predefined curriculum and serving as a unit for achieving a certain learning goal.

Adaptive OCW - OCW that is suitable to be used in adaptive learning environments.

Curriculum - structured plan that describes the educational program that is used in the educational resource [34].

Multimedia Presentation - a digital slide presentation which includes diverse media objects such as graphics and videos [22].

Learning Design (LD) - a sequence of (collaborative) learning activities. It can incorporate single learner content, but also collaborative tasks such as discussion, voting, small group debate, etc. LD can be stored, re-used, customized, etc [44].

3.3 Authoring

Cooperation - a division of the labor among participants, into activities where each person is responsible for a portion of the problem [45].

Collaboration - a mutual engagement of participants in a coordinated effort to solve a problem together [44].

Communities of Practice - the communities in which there exists "the sustained pursuit of shared enterprise" [56].

The Community-Build System (CBS) - a system for content creation by a community operating on a dedicated engine (e.g. Wiki) [47]. Also, it is a system of virtual collaborations organized to provide an open resource development environment within a given community.

Wiki - a Website that allows visitors to add, remove, edit and change content [13]. A collection of web pages designed to enable anyone who accesses it to contribute or modify content using a simplified markup language[4]. A web application which allows people to add, modify, or delete content in collaboration with others[5].

Semantic Wiki - a wiki that enables simple and quick collaborative text editing over the Web and Semantic Web. Semantic wiki extends a classical wiki by integrating it with the management capabilities for the formal knowledge representations[6].

Collaborative Authoring - occurs in project like settings, where the project delegates authoring sub-tasks to a group of authors. This kind of authoring needs synchronization, dialogue support, and coordination of the whole project [15].

Cooperative Authoring - mainly involves synchronous re-usage of authoring products, such as course materials, libraries, ontologies, etc [15].

Crowdsourcing - a problem-solving approach that outsources tasks to an undefined, often anonymous, population [24].

Workflow - a sequence of industrial, administrative, or other processes through which a piece of work passes from initiation to completion.

Workflow-based Approach - an approach that uses workflow for solving a task.

Social Annotation (Social Tagging, Folksonomy) - a system of classification derived from the practice and method of collaboratively creating and translating tags to annotate and categorize content [40].

Awareness - in the context of collaborative authoring, it is an understanding of activities of other collaborators, which provides a context for the own activity [20].

Knowledge Sharing - an activity where agents - individuals, communities, or organizations - exchange their knowledge - information, skills, or expertise [25].

3.4 Technical Concepts and Technologies

Ontology - a formal naming and definition of the types, properties, and inter-relationships of the entities that really or fundamentally exist for a particular domain of discourse[7].

Activity feed (Activity Stream) - a list of recent activities performed by an individual(s), typically on a single website or on a single content piece.

---

[4] http://wikipedia.com.
[5] http://en.wikipedia.org/wiki/Wiki.
[6] https://goranzugic.wordpress.com/2010/09/09/semantic-wikis/.
[7] http://en.wikipedia.org/wiki/Ontology_(information_science).

Mashup - (in the domain of OCW authoring) it is a web page, or web application, that uses content from more than one source to create a single new learning object or OCW.

RSS - originally *RDF Site Summary*, but often dubbed as *Really Simple Syndication*. It is a mechanism to publish a feed of frequently updated information: blog entries, news headlines, audio, video etc.[8].

RSS Aggregator - a tool that periodically checks for updates to the RSS feed and keeps the user informed of any changes [3].

Grid - a collection of independently owned and administered resources which have been joined together by a software and hardware infrastructure that interacts with the resources and the users of the resources to provide coordinated dynamic resource sharing in a dependable and consistent way according to policies that have been agreed to by all parties [21].

Virtual Learning Environment (VLE) - tools that support e-learning through integrated provision of learning materials, and communication, administration, and assessment tools [12].

4 Experimental Implementation of Collaborative OCW Authoring System

Based on the study conducted, we have noticed the lack of a complex approach dealing with the most of the challenging tasks. While there are systems available implementing one or another aspect of the collaborative OCW authoring, they usually ignore other aspects. In order to evaluate how the approaches we consider to be the most promising will work together, we have designed, developed and evaluated the collaborative OCW authoring platform SlideWiki.

4.1 Conceptual Design

As can be seen from the Table 4, most of the challenging tasks can be covered by five strategical approaches. Moreover, combining the approaches gives a synergistic effect. We have called the resulting complex solution *CrowdLearn concept* and illustrate it in Fig. 4. Below we discuss the five main components and their interrelations.

Crowdsourcing. There are already vast amounts of amateur and expert users which are collaborating and contributing on the Social Web. Harnessing the power of such crowds can significantly enhance and widen the distribution of e-learning content. Crowd-sourcing as a distributed problem-solving and production model is defined to address this aspect of collective intelligence [23]. CrowdLearn as its main innovation combines the crowd-sourcing techniques

---

[8] http://en.wikipedia.org/wiki/RSS.

Table 4. Chosen approaches for solving OCW collaborative authoring challenges

Aspect	Approach chosen	Remarks
Content co-creation		
Content authoring	Crowdsourcing, wiki	Content objects, metadata, assessment items and styles are reusable and fully versioned
Metadata authoring	Crowdsourcing, wiki	
Assessmnt. items auth	Crowdsourcing, wiki	
Quality assurance	Social networking, crowdsourcing, wiki, semantic structuring	Synergistic effect
Categorization	Crowdsourcing (social annotation)	
Personalization	Semantic structuring, crowdsourcing (social annotation)	
Localization	Crowdsourcing, wiki	See for more details
Reuse and re-purpose		
Search, filtering, aggregation	Standard compliance, crowdsourcing (social annotation), semantic structuring	
Remixing	Semantic structuring	Using WikiApp data model
Organization	Semantic structuring	Using WikiApp data model
Social Collaboration		
Negotiation	Social Networking	
Awareness	Social Networking	
Network building	Social Networking	
Engagement	Digital badges, Social Networking	

with the creation of highly-structured e-learning content. E-learning material when combined with crowd-sourcing and collaborative social approaches can help to cultivate innovation by collecting and expressing (contradicting) individual's ideas. As Paulo Freire wrote in his 1968 book *Pedagogy of the Oppressed*, 'Education must begin with the solution of the teacher-student contradiction, by reconciling the poles of the contradiction so that both are simultaneously teachers and students...'. Therefore, crowd-sourcing in the domain of educational material not only increases the amount of e-learning content but also improves the quality of the content. Our concept assumes application of crowdsourcing techniques to all kinds of the content and its metadata, including self-assessment items. This, together with social networking and semantic structuring, completely solves the challenges from the *content co-creation* aspect group. The content quality assurance is then reached through facilitation of small contributions from the crowd and ability to discuss the individual learning artifacts, such as an individual slide or a question. Social annotation of the content pieces serves as a basis for content categorization and customization. An important task our concept allows to be done by the crowd is content localization. Here the content structuring plays a crucial role in improving the quality of the translation, due to the possibility to translate and edit each learning artifact individually.

Wiki. The wiki paradigm supposes the crowdsourcing of the content supported by facilitation of small contributions, formatting and version control. To be able to deal with structured content, the wiki paradigm needs a more

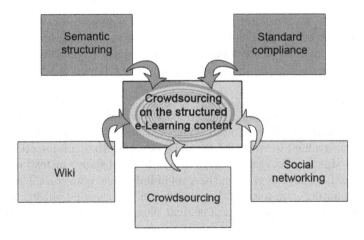

Fig. 4. CrowdLearn concept.

comprehensive data model. Our CrowdLearn concept assumes the content to be presented according to our previously developed WikiApp data model, described in details in [51]. The WikiApp data model is a refinement of the traditional entity-relationship data model. It adds some additional formalisms in order to make users as well as ownership, part-of and based-on relationships first-class citizens of the data model. A set of content objects connected by *part-of* relations can be arranged and manipulated in exactly the same manner, as an individual non-structured object. The model natively supports versioning and structuring of the different content objects.

The WikiApp model assumes that all content objects are versioned using the timestamp $c_{t,i}$ and the base content object relation $b_{t,i}$. In the spirit of the wiki paradigm, there is no deletion or updating of existing, versioned content objects. Instead new revisions of the content objects are created and linked to their base objects via the *base-content-object* relation. All operations have to be performed by a specific user and the newly created content objects will have this user being associated as their owner.

The model is compatible with *both* the relational data model and the *Resource Description Framework* (RDF) data model (i.e. it is straightforward to map it to each one of these). When implemented as a relational data, content types correspond to tables and content objects to rows in these tables. Functional attributes and relationships as well as the *owner* and *base-content-object* relationships can be modeled as columns (the latter three representing foreign-key relationships) in these tables. The implementation of the WikiApp model in RDF is slightly more straightforward: content types resemble classes and content objects instances of these classes. Attributes and relationships can be attached to the classes via `rdfs:domain` and `rdfs:range` definitions and directly used as properties of the respective instances.

Watching the users, as well as *following* the learning objects operations are natively supported by the model. This allows users to receive information about changes of the followed content object or new objects created by the watched user. Also, these operations allow to easily find the followed object or user.

Semantic structuring. The semantic structuring of the content is implemented via using WikiApp data model discussed in the paragraph above. Instead of dealing with large learning objects (often whole presentations or tests), we decompose them into fine-grained *learning artifacts*. Thus, rather than a large presentation, user will be able to edit, discuss and reuse individual slides; instead of a whole test she/he will be able to work on the level of individual questions. This concept efficiently facilitates the reuse and re-purpose of the learning objects. Semantic structuring facilitates application of content personalization mechanisms, allowing recommendation systems to work with a finer tuned setup. Semantic structuring together with standard compliance and social annotation facilitates content publishing, aggregation and filtering due to the ability of algorithms to work with finer grained content.

Social networking. The theoretical foundations for e-Learning 2.0 are drawn from *social constructivism* [54]. It is assumed that students learn as they work together to understand their experiences and create meaning. In this view, teachers are knowers who craft a curriculum to support a self-directed, collaborative search and discussion for meanings. Supporting social networking activities in CrowdLearn enables students to proactively interact with each other to acquire knowledge. With the CrowdLearn concept we address the following social networking activities:

- Users can follow individual learning objects as well as other users activities to receive notification messages about their updates.
- Users can discuss the content of learning objects in a forum-like manner.
- Users can share the learning objects within their social network websites such as Facebook, Google Plus, LinkedIn, etc.
- Users can rate the available questions in terms of their difficulty.

Besides increasing of the learning process quality, social activities improve the quality of the created learning material. Even when answering a quiz, users can contribute by analysing the quality of the questions and making suggestions of how to improve them. Thus, the knowledge is being created not only explicitly by contributors, but also implicitly through discussions, answering the questions of assessment tests, or in other words through native learning activities.

Standard-compliance. The costs associated with building high-quality e-learning content are high. One solution to decrease the costs is to author structured and reusable e-learning content that can be repurposed in different ways. To facilitate this, it should be possible to migrate content between different *Learning Management Systems* (LMSs). However, often content migration is not completely adequate and can thus result in loss of valuable content, meta-data or structure. Even if the transfer is possible, moving the content between systems can be more costly than just redeveloping that course in the new system. The strategy to overcome this challenge is the standard-compliance of both LMS and content. In that regard, we adopted the *SCORM standard* [2] and practical recommendations [1] and expanded the standard for the collaborative model.

4.2   Implementation

Data Model. Our SlideWiki example application uses two implementations of WikiApp data model. The first implementation is used for managing slides and presentations. It includes individual slides (consisting mainly of HTML snippets, SVG images and meta-data), decks (being ordered sequences of slides and sub-decks), themes (which are associated as default styles with decks and users) and media assets (which are used within slides). The second implementation was developed for managing questions and assessment tests. It includes questions for the slide material (the question is assigned to all slide revisions), tests (which could be organized manually by user or created automatically in accordance with the deck content), and answers (which are the part of the questions).

We implicitly connected these two WikiApp instances by adding two relations. Firstly, we assigned questions to slides. Thus, during the learning process users are able to answer the tests and have a look at the assigned slide if necessary. The important issue here is that we assign question not to individual slide revision, but for the slide itself. This decision gives an opportunity to create a new slide revision, that already has a list of questions, collected from other revisions. Secondly, we assigned assessment tests to concrete deck revisions. Thus the automatically created test saves the structure of the corresponding deck revision. This allows us to use module-based assessment to score the test results.

In order to publish the SlideWiki content in accessible way, we have adapted the developed data model for linked data. In order to do so, we first created an ontology for our WikiApp data model. While developing the ontology we used the approach and recommendations discussed in [53]. Especially, we focused on metrics from the interoperability dimension, carefully choosing the properties to reuse and providing high-quality documentation for developed classes and properties. Following the WikiApp data model formalization, the core of the ontology is built on three main classes: (1) *wa:Container* storing the content type properties, (2) *wa:ContentObject* for storing the content objects and all kinds of relations between them and (3) *wa:User*, a subclass of *foaf:Person*. Being an implementation of WikiApp data model, the ontology requires each content object to have only the *dcterms:created* property specified. This ensures high flexibility and interoperability of the ontology, making it easy to be reused in a wide range of applications.

Architecture and Technical Solutions. The SlideWiki application makes extensive use of the model-view-controller (MVC) architecture pattern. The MVC architecture enables the decoupling of the user interface, program logic and

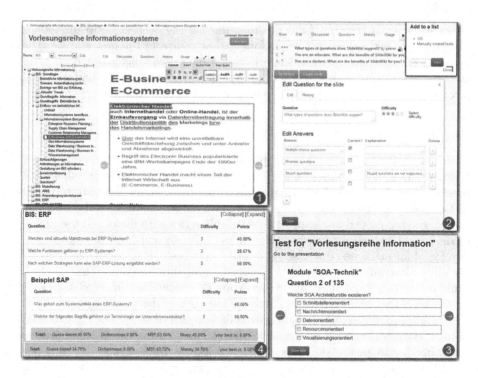

Fig. 5. Four screenshots of SlideWiki features. **1** - Tree structure of the presentation, inline WYSIWYG editor; **2** - Editing of a question and manual assigning to a test using lists; **3** - Question in learning mode with correct answers displayed; **4** - Module-based scoring of an assessment test.

database controllers and thus allows developers to maintain each of these components separately. The implementation comprises the main components: authoring, change management, import/export, linked data interface, e-assessment and translation. We briefly walk-through these components in the sequel.

Authoring. SlideWiki employs an inline HTML5 based WYSIWYG (What-You-See-Is-What-You-Get) text editor for authoring the presentation slides (cf. Fig. 5, image 1). Using this approach, users will see the slideshow output at the same time as they are authoring their slides. The editor is implemented based on ALOHA editor[9] extended with some additional features such as image manager, source manager, equation editor. The inline editor uses SVG images for drawing shapes on slide canvas. Editing SVG images is supported by SVG-edit[10] with some predefined shapes which are commonly used in presentations. For logical structuring of presentations, SlideWiki utilizes a tree structure in which users can append new or existing slides/decks and drag & drop items for positioning.

---

[9] http://aloha-editor.org/.
[10] http://code.google.com/p/svg-edit/.

When creating presentation decks, users can assign appropriate tags as well as footer text, default theme/transition, abstract and additional meta-data to the deck.

Change management. Revision control is natively supported by WikiApp data model. We just define rules and restrictions to increase the performance. There are different circumstances in SlideWiki for which new slide or deck revisions have to be created.

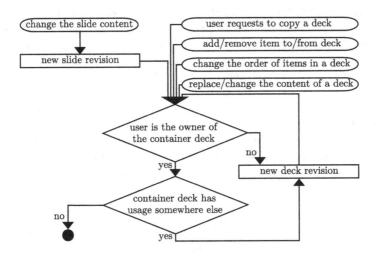

Fig. 6. Decision flow during the creation of new slide and deck revisions.

For decks, however, the situation is slightly more complicated, since we wanted to avoid an uncontrolled proliferation of deck revisions. This would, however, happen due to the fact, that every change of a slide would also trigger the creation of a new deck revision for all the decks the slide is a part of. Hence, we follow a more retentive strategy. We identified three situations which have to cause the creation of new revisions:

- The user specifically requests to create a new deck revision.
- The content of a deck is modified (e.g. slide order is changed, change in slides content, adding or deleting slides to/from the deck, replacing a deck content with new content, etc.) by a user which is neither the owner of a deck nor a member of the deck's editor group.
- The content of a deck is modified by the owner of a deck but the deck is used somewhere else.

The decision flow is presented in Fig. 6. In addition, when creating a new deck revision, we always need to recursively spread the change into the parent decks and create new revisions for them if necessary.

Import/Export. SlideWiki implementation addresses *interoperability* as its first class citizen. As shown in Fig. 7, SlideWiki supports import/export of the content from/to existing desktop applications and LORs thereby allowing users from other LMSs to access the created content. The main data format used in SlideWiki is HTML. However, there are other popular presentation formats commonly used by desktop application users, such as PowerPoint .pptx presentations, LaTeX and others. We implemented import of the slides from .pptx format and work on the LaTeX format support is in progress.

E-Assessment. SlideWiki supports the creation of questions and self-assessment tests based on slide material. Each question has to be assigned to at least one slide. Important note here, that the question is assigned not to the slide revision, but to slide itself. Thus, when a new slide revision appears, it continues to include all the list of previously assigned questions. Questions can be combined into tests. The *automatically created* tests include the last question revisions from all the slides within the current deck revision. Manually created tests present a collection of chosen questions and currently cannot be manipulated as objects (cf. Fig. 5, image 2). Thus, in our implementation only questions and answers have to be placed under the version control. However, their structure is trivial and the logic of creating their new revisions is intuitive. We just restricted the

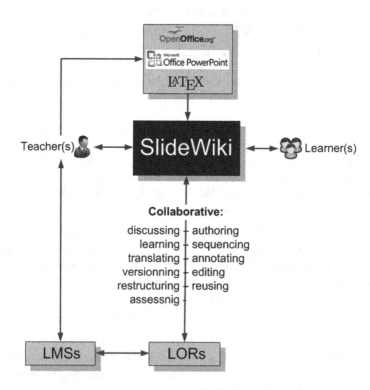

Fig. 7. SlideWiki interoperability scheme.

number of new revisions to be created similarly with the decks: changes made by the question owner do not trigger a new revision creation. For now, only multiple-choice (and multiple-mark) question type is implemented, however in the future we plan to expand the list of supported types.

Multilinguality. The implementation of multilingual content support in Slide Wiki is based on the co-evolution paradigm described in details in [52]. The implementation of co-evolution of source object content and its translations supposes the implementation of three operations: (1) initial translation, (2) synchronization and (3) merging of the revisions (Fig. 8).

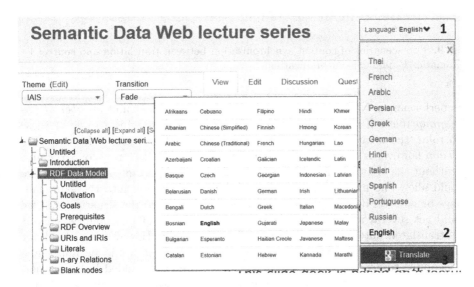

Fig. 8. Interface of translations management: 1 - language of selected object; 2 - a drop-down list with links to languages available fro the selected object; 3 - button for translation; 4 - dynamically updated list of the languages supported by Google Translate service

Our architecture allowed us to implement a *translation* operation backed by the Google Translate service. After translation into one of 71 currently supported languages, the presentation can be edited, re-structured and reused independently from its source.

To enable *synchronization* of original and translated versions, every further revision of translated objects inherits the link to the source revision (see v2.1 at Fig. 9). The changes in the original version of the object cause the creation of new revision v1.1. Additionally, users are notified of translations that have become out of sync with the source (exclamation marks in v2.0 and v2.1).

SlideWiki implements the revision control in accordance with the WikiApp data model, where *merging the revisions* is supported as one of the core operations. However, as discussed above we defined rules and restrictions to increase

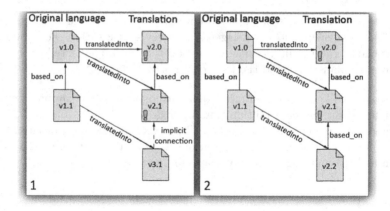

Fig. 9. Two scenarios of content synchronization between translation and source: 1 - automatic; 2 - manual synchronization

the performance. Namely, we introduced the *content owner* and *member of editor group* roles. If the changes are made by a user belonging to one of these two roles, the creation of a new *deck revision* is not triggered (the new *slide revision* however is created). As we allow the owner of a deck revision to change it without the creation of a new revision, it was an important issue whether we should allow the multiple translation of the same revision into the same language or not. We decided to allow it, however, this led to the situation that we would get several identical presentations with content of bad quality, since it was translated automatically and not edited manually. However, we could not disable the multiple translations, because in that case it would be impossible for example to get translations of new slides if they were added by the owner. Thus, merging the revisions became the crucial operation, not only for merging back-translation with the source, but also for merging multiple translations in the same language.

4.3   Evaluation

To evaluate the real-life usability of SlideWiki, we used it for accompanying an information systems lecture at Chemnitz Technical University. We structured the slides within the lecture series and added questions for student self-assessment before the final exam. We informed them about the different e-learning features of SlideWiki, in particular, how to prepare for the exam using SlideWiki. The experiment was not obligatory but students actively contributed by creating additional questions and fixing mistakes. The experiment was announced to 30 students of the second year and 28 of them registered at SlideWiki.

The students were working with SlideWiki for several weeks, and we collected the statistics for that period. During that period, they created 252 new slide revisions which some of them were totally new slides, others were improved versions of the original lecture slides. Originally the whole course had 130 questions, and students changed 13 of them, fixing the typos or adding additional distractors to multiple choice questions. In total, students performed 287 self-assessment tests. The majority of these used the automatically and randomly created tests covering the whole course material. 20 tests included only difficult questions, 2 asked to show the questions with increasing difficulty. This showed us that the students liked the diversity of test organization. Students also liked the possibility to limit the number of questions – 80 attempts were made with such a setting. 8 students reached the 100 % result for the whole course. On average, it took them 6 attempts before they succeeded.

After the experiment we can claim, that more active SlideWiki users received better marks on the real examination. It shows that SlideWiki not only allows students to prepare for the examinations, but also engages them in active participation that helps to improve the quality of the learning. After the end of the semester, we asked the participants to fill out a questionnaire which consisted of three parts: usability experience questions, learning quality questions and open questions for collecting the qualitative feedbacks. We collected 9 questionnaires that were filled out completely. They show us emergent problems and directions for the future.

In the first part of the questionnaire we included questions recommended by *System Usability Scale* (SUS) [30] system to grade the usability of SlideWiki. SUS is a standardized, simple, ten-item Likert scale-based questionnaire[11] giving a global view of subjective assessments of usability. It yields a single number in the range of 0 to 100 which represents a composite measure of the overall usability of the system. The results of our survey showed a mean usability score of 67.2 for SlideWiki which indicates a reasonable level of usability.

The second part of the questionnaire aimed to determine whether the SlideWiki helps to improve the quality of learning. It consisted of four questions with five options from "absolutely agree (1)" to "absolutely disagree (5)". The evaluation results for these two parts are presented in Fig. 10.

Although the positive answers prevail, we were not satisfied by the fact that for many questions a third of participants chose the neutral value. It could be a signal, that students do not completely understand the question or are not 100 % sure about the result.The last part of the questionnaire helped us to understand the reasons. We included four open questions:

1. What did you like most about Slidewiki?
2. What did you like least about Slidewiki?
3. What can we do to improve the Slidewiki's usability?
4. What features would you add to Slidewiki?

Within the answers we found repeated complaints about several bugs, that interfered the working process. We consider this fact to be the main reason of neutral and contradictory values. However, we collected also positive opinions, especially about features and possibilities that SlideWiki allows. Three of the recipients mentioned that they mostly liked that SlideWiki is easy to use, four

---

[11] www.usabilitynet.org/trump/documents/Suschapt.doc.

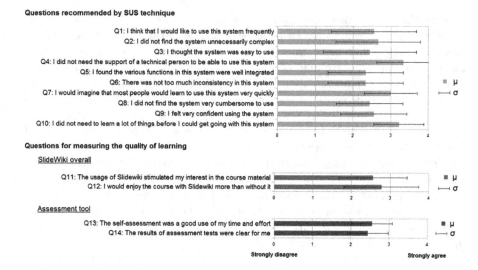

Fig. 10. Results of SlideWiki evaluation survey: mean μ and standard deviation σ.

of them noted, that they liked the idea of collaborative work and sharing the presentations itself. Within the collected answers we also got important suggestions, which could be roughly divided into two groups:

- Suggestions about desired improvements of existing features such as displaying the test results graphically, supporting more import formats, improving the SVG editor etc.
- Suggestions about totally new features, several of those were later implemented, e.g. translation, templates for presentation structure, etc.

Also we collected a few suggestions about features that were already implemented, but users were not aware of them. This encourages us to improve the documentation as well as to enhance the simplicity and clearness of the user interface. One of the students drew our attention to security issues.

The results of our evaluation showed that our concept is clear to the users, they like this way of learning, storing and sharing of the presentations. However, we need to improve the user interface, fix minor bugs and spend more effort on privacy and security issues.

5   Conclusions and Future Work

In the paper we presented the analysis of collaborative authoring of OCW state-of-art. We have answered the following research questions:

1. *What are the main challenging tasks in collaborative authoring of reusable OCW?*
2. *Which technologies are being used to solve these tasks?*

3. *How well covered are the challenging tasks in the scientific literature?*
4. *What are the main gaps in state-of-art research and what are the most promising areas of future work in the field?*

According to our study, the main challenging tasks of collaborative authoring of OCW can be divided in three dimensions: (1)content co-creation, including metadata and assessment items authoring, classification, localization, personalization and quality assurance; (2) content reuse and re-purpose, including content organization, remixing, search, aggregation and filtering; (3) social collaboration, including user coordination and user engagement. The technologies currently being used are mostly known as Web 2.0 technologies and Artificial Intelligence. We summarize the answers for these two questions in the mind map presented in Fig. 3. The issues and technologies are covered in the literature but lacking in depth, that means there are not enough *detailed* and *focused* studies on specific aspects and technologies This is especially true for some of them (workflow-based collaborative authoring, social games for user engagement, content localization, content remixing, network building).

Based on the study results, we have developed a collaborative OCW authoring platform SlideWiki - a social web-based application targeting slide presentations and e-assessments. According to the evaluation, the platform implementation is promising, although it needs further development. Beside the usability improvements, our first direction for future work is to implement a completely SCORM-compliant LMS and authoring tool, based on SlideWiki. This will allow us to exchange the content with other SCORM-compliant LMSs. Also, in a real e-learning scenario, learners come from different environments, have different ages and educational backgrounds. These heterogeneity in user profiles is crucial to be addressed when enhancing the CrowdLearn concept. New approaches should provide the possibility to *personalize* the learning process. Thus, our second direction is providing personalized content based on initial user assessment. The third direction for future work is to support the annotation of learning objects using standard metadata schemes. We aim to implement the $LRMI^{12}$ metadata schemes to facilitate end-user search and discovery of educational resources.

References

1. SCORM Users guide for programmers. Technical report, ADL (2011). URL http://www.adlnet.gov/wp-content/uploads/2011/12/SCORM_Users_Guide_for_Programmers.pdf
2. SCORM 2004 4th Edition Specification. Technical report, ADL (2011). URL http://www.adlnet.gov/wp-content/uploads/2011/07/SCORM_2004_4ED_v1_1_Doc_Suite.zip
3. Andersen, P.: What is Web 2.0?: ideas, technologies and implications for education, vol. 1. JISC Bristol, UK (2007)

---

[12] Learning Resource Metadata Initiative: www.lrmi.net/.

4. Auinger, A., Ebner, M., Nedbal, D., Holzinger, A.: Mixing content and end-less collaboration – Mashups: towards future personal learning environments. In: Stephanidis, C. (ed.) UAHCI 2009, Part III. LNCS, vol. 5616, pp. 14–23. Springer, Heidelberg (2009)

5. Bafoutsou, G., Mentzas, G.: Review and functional classification of collaborative systems. Int. J. Inf. Manage. **22**(4), 281–305 (2002)

6. Blau, I: E-collaboration within, between, and without institutions: Towards better functioning of online groups through networks. In: Interdisciplinary Applications of Electronic Collaboration Approaches and Technologies, p. 188 (2013)

7. Brusilovsky, P.: Adaptive hypermedia. User Model. User-Adap. Inter. **11**(1–2), 87–110 (2001)

8. Brusilovsky, P.: Developing adaptive educational hypermedia systems: from design models to authoring tools. In: Murray, T., Blessing, S.B., Ainsworth, S. (eds.) Authoring Tools for Advanced Technology Learning Environments, pp. 377–409. Springer, The Netherlands (2003)

9. Cai, J., Zhou, Y., Ruan, G.: Learning activities design of a blended learning course based on public internet. In: 2010 2nd International Conference on Education Technology and Computer (ICETC), vol. 3, pp. V3-297–V3-299. IEEE (2010)

10. Chiribuca, D., Hunyadi, D., Popa, E.M.: The educational semantic web. In: Mastorakis, N.E., Demiralp, M., Mladenov, V., Bojkovic, Z. (eds.) Proceedings of the 8th Conference on Applied Informatics and Communications (AIC 2008), World Scientific and Engineering Academy and Society (WSEAS), Stevens Point, Wisconsin, USA, pp. 314–319 (2008)

11. Claros, I., Cobos, R.: Social media learning: an approach for composition of multimedia interactive object in a collaborative learning environment. In: 2013 IEEE 17th International Conference on Computer Supported Cooperative Work in Design (CSCWD), pp. 570–575. IEEE (2013)

12. Currier, S.: Integrating information resources and online learning in the UK. In: Proceedings of the International Conference on Computers in Education, pp. 818–822. IEEE (2002)

13. Dahlan, H.M., Ali, I.M., Hussin, A.R.C.: Suitability of collaborative learning activities in web 2.0 environment. In: 2010 International Conference on User Science and Engineering (i-USEr), pp. 65–70. IEEE (2010)

14. De Maio, C., Fenza, G., Loia, V., Senatore, S., Gaeta, M., Orciuoli, F.: RSS-generated contents through personalizing e-learning agents. In: Ninth International Conference on Intelligent Systems Design and Applications, ISDA 2009, pp. 49–54. IEEE (2009)

15. Dicheva, D., Aroyo, L., Cristea, A.I.: Collaborative courseware authoring support. CATE (2002)

16. DiNucci, D.: Design & new media: fragmented future-web development faces a process of mitosis, mutation, and natural selection. Print N.Y. **53**, 32–35 (1999)

17. Dyba, T., Dingsoyr, T., Hanssen, G.K.: Applying systematic reviews to diverse study types: an experience report. In: Proceedings of the First International Symposium on Empirical Software Engineering and Measurement, ESEM 2007, pp. 225–234. IEEE Computer Society, Washington, DC (2007). ISBN 0-7695-2886-4. doi:http://dx.doi.org/10.1109/ESEM.2007.21

18. García, F.J., García, J.: Educational hypermedia resources facilitator. Comput. Educ. **44**(3), 301–325 (2005)

19. Glover, I., Zhijie, X., Hardaker, G.: Online annotation-research and practices. Comput. Educ. **49**(4), 1308–1320 (2007)

20. Halimi, K., Seridi, H., Faron-Zucker, C.: Solearn: a social learning network. In: 2011 International Conference on Computational Aspects of Social Networks (CASoN), pp. 130–135. IEEE (2011)

21. Hamid, A., Tiawa, D., Islam, M., Alias, N., Omar, A.H., et al.: An efficient authoring activities infrastructure design through grid portal technology. In: The 7th WSEAS International Conference on Engineering Education, pp. 146–151 (2010)

22. Hong, C., Kim, Y.: The multimedia authoring in collaborative e-learning system. In: 2012 6th International Conference on New Trends in Information Science and Service Science and Data Mining (ISSDM), pp. 158–161. IEEE (2012)

23. Howe, J.: The rise of crowdsourcing. Wired Mag. **14**(6), 6 (2006). http://www.wired.com/wired/archive/14.06/crowds.html

24. Howe, J.: The rise of crowdsourcing. Wired Mag. **14**(6), 1–4 (2006)

25. Ireson, N., Burel, G.: Knowledge sharing in e-collaboration. In: Wimmer, M.A., Chappelet, J.-L., Janssen, M., Scholl, H.J. (eds.) EGOV 2010. LNCS, vol. 6228, pp. 351–362. Springer, Heidelberg (2010)

26. Johnson, C.M.: A survey of current research on online communities of practice. Internet High. Educ. **4**(1), 45–60 (2001). doi:10.1016/S1096-7516(01)00047-1. ISSN 1096-7516. http://www.sciencedirect.com/science/article/pii/S1096751601000471

27. Kitchenham, B.: Procedures for performing systematic reviews. Technical report, Keele University and NICTA (2004)

28. Lee, B., Ge, S.: Personalisation and sociability of open knowledge management based on social tagging. Online Inf. Rev. **34**(4), 618–625 (2010)

29. Leinonen, T., Purma, J., Poldoja, H., Toikkanen, T.: Information architecture and design solutions scaffolding authoring of open educational resources. IEEE Trans. Learn. Technol. **3**(2), 116–128 (2010)

30. Lewis, J.R., Sauro, J.: The factor structure of the system usability scale. In: Kurosu, M. (ed.) HCD 2009. LNCS, vol. 5619, pp. 94–103. Springer, Heidelberg (2009). doi:10.1007/978-3-642-02806-9_12. ISBN 978-3-642-02805-2. URL http://dx.doi.org/10.1007/978-3-642-02806-9_12

31. Li, Y., Dong, M., Huang, R.: Developing a collaborative e-learning environment based upon semantic wiki: from design models to application scenarios. In: 2010 IEEE 10th International Conference on Advanced Learning Technologies (ICALT), pp. 222–226. IEEE (2010)

32. IEEE LTSC: Learning object definition. Technical report, LTSC Learning Technology Standards Committee of the Institute of Electrical and Electronics Engineers (IEEE) (2002)

33. Luo, A., Ng'ambi, D., Hanss, T.: Towards building a productive, scalable and sustainable collaboration model for open educational resources. In: Proceedings of the 16th ACM International Conference on Supporting Group Work, pp. 273–282. ACM (2010)

34. Marcilla, S., et al.: Framework for the development and reuse of educational resources in the area of information systems management for different degrees at a technical university. In: Proceedings of the 2012 IEEE Global Engineering Education Conference (EDUCON), pp. 1–9 (2012)

35. Moxley, J.: Datagogies, writing spaces, and the age of peer production. Comput. Compos. **25**(2), 182–202 (2008)

36. Mu, S., Zhang, X., Zuo, P.: Research on the construction of open education resources based on semantic wiki. In: Cheung, S.K.S., Fong, J., Kwok, L.-F., Li, K., Kwan, R. (eds.) ICHL 2012. LNCS, vol. 7411, pp. 283–293. Springer, Heidelberg (2012)

37. Nesic, S., Jazayeri, M., Landoni, M., Gasevic, D.: Using semantic documents, social networking in authoring of course material: an empirical study. In: 2010 IEEE 10th International Conference on Advanced Learning Technologies (ICALT), pp. 666–670. IEEE (2010)

38. Ola, A.G., Bada, A.O., Omojokun, E., Adekoya, A.: Actualizing learning and teaching best practices in online education with open architecture and standards. In: Proceedings of International Conference on Information Security and Privacy, pp. 1790–2511 (2009)

39. Pansanato, L.T.E., Fortes, R.P.M.: System description: an orienteering strategy to browse semantically-enhanced educational wiki pages. In: Franconi, E., Kifer, M., May, W. (eds.) ESWC 2007. LNCS, vol. 4519, pp. 809–818. Springer, Heidelberg (2007)

40. Peters, I.: Folksonomies: Indexing and Retrieval in Web 2.0, vol. 1. Walter de Gruyter, Berlin (2009)

41. Porcello, D., Hsi, S.: Crowdsourcing and curating online education resources. Science **341**(6143), 240–241 (2013)

42. Posner, I.R., Baecker, R.M.: How people write together [groupware]. In: 1992 Proceedings of the Twenty-Fifth Hawaii International Conference on System Sciences, vol. 4, pp. 127–138. IEEE (1992)

43. Rich, P.J., Hannafin, M.: Video annotation tools technologies to scaffold, structure, and transform teacher reflection. J. Teacher Educ. **60**(1), 52–67 (2009)

44. Romero-Moreno, L.M., Javier Ortega, F., Troyano, J.A.: Obtaining adaptation of virtual courses by using a collaborative tool and learning design. In: 2007 Proceedings of the Euro American Conference on Telematics and Information Systems, p. 42. ACM (2007)

45. Roschelle, J., Teasley, S.D.: The construction of shared knowledge in collaborative problem solving. In: O'Malley, C. (ed.) Computer Supported Collaborative Learning, pp. 69–97. Springer, Heidelberg (1995)

46. Rosmala, D., et al.: Study of social networking usage in higher education environment. Procedia Soc. Behav. Sci. **67**, 156–166 (2012)

47. Różewski, P.: Model of community-build system for knowledge development. In: Jędrzejowicz, P., Nguyen, N.T., Hoang, K. (eds.) ICCCI 2011, Part II. LNCS, vol. 6923, pp. 50–59. Springer, Heidelberg (2011)

48. Shepherd, C.: Objects of interest. Brighton East Sussex: Fastrak Consulting Limited (2000). Accessed 14 June 2005

49. Smith, K.A.: Cooperative learning: effective teamwork for engineering classrooms. In: 1995 IEEE Frontiers in Education Conference (FIE), vol. 1, pp. 2b5-13–2b5-18. IEEE (1995)

50. Strobel, J., Jonassen, D.H., Ionas, I.G.: The evolution of a collaborative authoring system for non-linear hypertext: a design-based research study. Comput. Educ. **51**(1), 67–85 (2008)

51. Tarasowa, D., Khalili, A., Auer, S., Unbehauen, J.: Crowdlearn: crowd-sourcing the creation of highly-structured e-learning content. In: CSEDU, pp. 33–42 (2013)

52. Tarasowa, D., Auer, S., Khalili, A., Unbehauen, J.: Crowd-sourcing (semantically) structured multilingual educational content (CoSMEC). Open Prax. **6**(2), 159–170 (2014)

53. Tarasowa, D., Lange, C., Auer, S.: Measuring the quality of relational-to-RDF mappings. In: Klinov, P., et al. (eds.) KESW 2015. CCIS, vol. 518, pp. 210–224. Springer, Heidelberg (2015). doi:10.1007/978-3-319-24543-0_16

54. Wang, X., Love, P.E.D., Klinc, R., Kim, M.J., Davis, P.R.: Integration of e-learning 2.0 with web 2.0. ITcon - Special Issue eLearning 2.0: Web 2.0-Based Soc. Learn. Built Environ. **17**, 387–396 (2012)
55. Wang, Z.-H.: The study of the educational resources sharing model based on grids. In: 2010 2nd International Workshop on Database Technology and Applications (DBTA), pp. 1–4. IEEE (2010)
56. Wenger, E.: Communities of Practice: Learning, Meaning, and Identity. Cambridge University Press, Cambridge (1998)

Teaching (with) Open and Linked Data

Teaching Linked Open Data Using Open Educational Resources

Alexander Mikroyannidis[1(✉)], John Domingue[1], Maria Maleshkova[2],
Barry Norton[3], and Elena Simperl[4]

[1] Knowledge Media Institute, The Open University, Milton Keynes, UK
{Alexander.Mikroyannidis,John.Domingue}@open.ac.uk
[2] Karlsruhe Institute of Technology, Karlsruhe, Germany
maria.maleshkova@kit.edu
[3] British Museum, London, UK
BNorton@britishmuseum.org
[4] University of Southampton, Southampton, UK
E.Simperl@soton.ac.uk

Abstract. Recent trends in online education have seen the emergence of Open Educational Resources (OERs) and Massive Open Online Courses (MOOCs) as an answer to the needs of learners and educators for open and reusable educational material, freely available on the web. At the same time, Big Data and the new analytics and business intelligence opportunities that they offer are creating a growing demand for data scientists possessing skills and detailed knowledge in this area. This chapter presents a methodology for the design and implementation of an educational curriculum about Linked Open Data, supported by multimodal OERs. These OERs have been implemented as a combination of living learning materials and activities (eBook, online courses, webinars, face-to-face training), produced via a rigorous process and validated by the data science community through continuous feedback.

Keywords: Open educational resources · Massive open online courses · Linked open data · Big data · Ebooks · Data science

1 Introduction

There is a revolution occurring now in higher education, largely driven by the availability of high quality online materials, also known as Open Educational Resources (OERs). OERs can be described as "teaching, learning and research resources that reside in the public domain or have been released under an intellectual property license that permits their free use or repurposing by others depending on which Creative Commons license is used" [1]. The emergence of OERs has greatly facilitated online education through the use and sharing of open and reusable learning resources on the web. Learners and educators can now access, download, remix, and republish a wide variety of quality learning materials available through open services provided in the cloud.

The OER movement aims in developing a comprehensive set of resources and content that is freely accessible and can be modified by anyone, whilst giving the

© Springer International Publishing Switzerland 2016
D. Mouromtsev and M. d'Aquin (Eds.): Open Data for Education, LNCS 9500, pp. 135–152, 2016.
DOI: 10.1007/978-3-319-30493-9_7

original author credit. OERs can comprise of any kind of learning resource material, textbook, papers, pictures, or web resources that is published in a format that can be copied or modified by anyone under a common licence. This very broad concept includes curriculum materials, educational software, computer-based learning systems, educational games, and more [9, 14].

The OER movement has appealed to a broad range of institutions, universities, researchers, teachers and scientists, who aim in opening up access to the world's knowledge resources [2, 16, 18]. Their mission is to freely distribute teaching materials of high quality into the public domain. Such OERs can then be customised, improved and shared with local communities. Additionally, they can be adapted for local and cultural contexts, such as language, level of study, pre-requirements, and learning outcomes [19].

The OER initiative has recently culminated in MOOCs (Massive Open Online Courses), which offer large numbers of students the opportunity to study high quality courses with prestigious universities. These initiatives have led to widespread publicity and also strategic dialogue in the higher education sector. The consensus within higher education is that after the Internet-induced revolutions in communication, business, entertainment, media, amongst others, it is now the turn of universities. Exactly where this revolution will lead is not yet known but some radical predictions have been made including the end of the need for university campuses [6].

At the same time, more and more industry sectors are in need of innovative data management services, creating a demand for data scientists possessing skills and detailed knowledge in this area. Declared by Harvard Business Review as the "sexiest job of the 21st century" [7], data science skills are becoming a key asset in any organization confronted with the daunting challenge of making sense of information that comes in varieties and volumes never encountered before. Ensuring the availability of such expertise will prove crucial if businesses are to reap the full benefits of these advanced data management technologies, and the know-how accumulated over the past years by researchers, technology enthusiasts and early adopters.

Linked Open Data (LOD) [4] has established itself as the de facto means for the publication of structured data over the web, enjoying amazing growth in terms of the number of organizations committing to use its core principles for exposing and interlinking data for seamless exchange, integration, and reuse [5]. More recently, data explosion on the web, fuelled by social networking, micro-blogging, as well as crowdsourcing, has led to the *Big Data* phenomenon [11, 12]. This is characterized by increasing volumes of structured, semi-structured and unstructured data, originating from sources that generate them at an increasing rate. This wealth of data provides numerous new analytic and business intelligence opportunities to various industry sectors.

The data scientist job title has been around for a while now, after being first introduced in 2008 to refer to the leads of data analytics efforts at two prominent IT companies in Silicon Valley [7]. It is typically linked to a number of core areas of expertise, from the ability to operate high-performance computing clusters and cloud-based infrastructures, to the know-how that is required to devise and apply sophisticated Big Data analytics techniques, and the creativity involved in designing powerful visualizations [10]. Moving further away from the purely technical, organizations are more and more looking into novel ways to capitalize on the data they own [3] and to generate added value from an increasing number of data sources openly available on the Web, a trend which has been coined as "open data" [15]. To do so, they need their employees to understand the legal and economic aspects of data-driven business development, as a prerequisite for the creation of product and services that turn open and corporate data assets into decision-making insight and commercial value.

Surviving in the data economy depends on hiring data professionals who master both the technical and non-technical facets of data science, from Big Data technology and data-driven storytelling, to new data monetization and innovation models. The challenge for managers is thus to identify and prioritize their knowledge gaps in this rapidly evolving, and to some extent interdisciplinary field, secure new talent, and train their existing staff into becoming proficient data practitioners and entrepreneurs.

Data scientists are, however, still a rare breed. Beyond the occasional data-centric start-up and the data analytics department of large corporations, the skills scarcity is already becoming a threat for many European companies and public sector organisations as they struggle to seize Big Data opportunities in a globalised world. A well-known McKinsey study [11] estimated already in 2011 that the United States will soon require 60 % more graduates able to handle large amounts of data as part of their daily jobs. With an economy of comparable size (by GDP) and growth prospects, Europe will most likely be confronted with a similar talent shortage of hundreds of thousands of qualified data scientists, and an even greater need of executives and support staff with basic data literacy. The number of job descriptions and an increasing demand in higher-education programs and professional training confirm this trend [8], with some EU countries forecasting an increase of almost 100 % in the demand for data science positions in less than a decade [13].

Combining these two trends in online education and data science, we have developed a methodology for building a LOD curriculum, implementing it via a rigorous production process and delivering it to the community of data scientists using a wide range of OERs. The following sections of this chapter describe this work in more detail. Section 2 introduces our approach in building a LOD curriculum tailored to the needs of data scientists. Section 3 presents the structure of the developed LOD curriculum. Section 4 discusses the methodology for the implementation and delivery of our curriculum as multimodal OERs. Finally, Sect. 5 presents the best practices we have distilled from the development and delivery of the LOD curriculum.

2   Building a LOD Curriculum

In the context of the European project EUCLID (Educational Curriculum for the Usage of Linked Data)[1], we have developed a comprehensive educational curriculum, supported by multimodal OERs and highly visible eLearning distribution channels. The EUCLID curriculum focuses on techniques and software to integrate, query, and visualize LOD, as core areas in which practitioners state to require most assistance.

---

[1] http://www.euclid-project.eu.

A significant part of the learning materials produced in the project consists of examples referring to real-world data sets and application scenarios, code snippets and demos that developers can run on their machines, as well as best practices and how-tos.

The EUCLID educational curriculum consists of a series of modules, each containing multimodal OERs, such as presentations, webinars, screencasts, exercises, eBook chapters, and online courses. These learning materials complement each other and are connected to deliver a comprehensive and concise training programme to the community. Learners are guided through these materials by following learning pathways, which are sequences of learning resources structured appropriately for achieving specific learning goals. Different types of eLearning distribution channels are targeted by each type of learning materials, including Apple and Android tablets, Amazon Kindles, as well as standard web browsers (see Fig. 1). All the EUCLID learning materials have been made available under a Creative Commons Attribution 3.0 Unported License[2]. This means that they can be shared, remixed, republished, as well as used for commercial purposes.

Instead of mock LOD examples, we have used in our learning materials and exercises a collection of datasets and tools that are deployed and used in real life. In particular, we use a number of large datasets including, for example, the MusicBrainz dataset[3], which contains 100Ms of triples. Our collection of tools includes the Information Workbench[4], Seevl[5], Sesame[6], OpenRefine[7] and GateCloud[8], all of which are used in real-life contexts. We also showcase scalable solutions, based upon industrial-strength repositories and automatic translations, e.g. by using the W3C standard R2RML[9] for generating RDF from large data contained in standard databases.

Additionally, the EUCLID project has had a strong focus on the community and has encouraged community engagement in the production of OERs through, for example, collecting user feedback via webinars, Twitter, LinkedIn, and more. We have combined online and real-world presence, and attempted to integrate with on-going activities in each sphere, such as mailing lists and wikis. The project has engaged with the LOD community, both practitioners and academics, by collecting user requirements as well as feedback to the OERs so that they can be tailored to what the learner really needs.

3   The EUCLID LOD Curriculum

The main target audience of this curriculum is data practitioners and professionals, who already use or aim to adopt LOD as means for publishing and accessing structured data over the Web. This has motivated the practical orientation of the learning materials

---

[2] http://creativecommons.org/licenses/by/3.0.

[3] https://musicbrainz.org/doc/MusicBrainz_Database.

[4] http://www.fluidops.com/en/portfolio/information_workbench.

[5] https://developer.seevl.fm.

[6] https://bitbucket.org/openrdf/sesame.

[7] http://openrefine.org.

[8] https://gatecloud.net.

[9] http://www.w3.org/TR/r2rml.

Fig. 1. A selection of EUCLID learning materials in different formats and platforms, i.e. eBooks and online courses for the web and the iPad.

and the use of directly appreciable examples. It is important to point out that the EUCLID curriculum focuses on providing real business application examples and relating the learnt topics to those examples. In particular, the goal is to provide a curriculum that includes use cases and scenarios, which directly demonstrate the practical applicability of the learned concepts and technologies.

The EUCLID LOD curriculum has been designed to gradually build up the learner's knowledge. It enables learners with previous knowledge on a specific area of interest to only briefly go over the introductory materials and directly dig into one of the more advanced modules. As shown in Fig. 2, the curriculum is composed of 6 modules that cover all the major aspects of the LOD consumption lifecycle. In terms of the targeted skills and knowledge that are to be gained, the curriculum provides three main levels of expertise:

- *Introductory level*– This level communicates the fundamental skills that are required in order to begin applying LOD technologies.
- *Intermediate level*– This level deals with more advanced topics, also specialising in different areas such as visualisation and query processing.

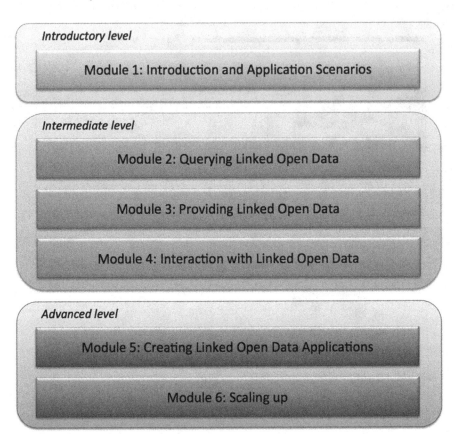

Fig. 2. The EUCLID LOD curriculum

- *Advanced level–* This level aims to provide expertise knowledge that is more specific to an area of use and requires somewhat extensive prior knowledge.

The 6 EUCLID modules have been structured to cover the following range of topics:

- *Module 1: Introduction and Application Scenarios.* This module introduces the main principles of LOD, the underlying technologies and background standards. It provides basic knowledge for how data can be published over the Web, how it can be queried, and what are the possible use cases and benefits. As an example, we use the development of a music portal (based on the MusicBrainz dataset), which facilitates access to a wide range of information and multimedia resources relating to music. The module also includes some multiple-choice questions in the form of a quiz, screencasts of popular tools and embedded videos.
- *Module 2: Querying LOD.* This module looks in detail at SPARQL (SPARQL Protocol and RDF Query Language) and introduces approaches for querying and updating semantic data. It covers the SPARQL algebra, the SPARQL protocol,

and provides examples for reasoning over LOD. The module uses examples from
the music domain, which can be directly tried out and ran over the MusicBrainz
dataset. This includes gaining some familiarity with the RDFS and OWL languages,
which allow developers to formulate generic and conceptual knowledge that can be
exploited by automatic reasoning services in order to enhance the power of
querying.

- *Module 3: Providing LOD.* This module covers the whole spectrum of LOD production and exposure. After a grounding in the LOD principles and best practices,
  with special emphasis on the VoID vocabulary, we cover R2RML, operating on
  relational databases, Open Refine, operating on spreadsheets, and GATECloud,
  operating on natural language. Finally, we describe the means to increase inter-
  linkage between datasets, especially the use of tools like Silk.
- *Module 4: Interaction with LOD.* This module focuses on providing means for
  exploring LOD. In particular, it gives an overview of current visualization tools and
  techniques, looking at semantic browsers and applications for presenting the data to
  the end used. We also describe existing search options, including faceted search,
  concept-based search and hybrid search, based on a mix of using semantic information and text processing. Finally, we conclude with approaches for LOD analysis, describing how available data can be synthesized and processed in order to
  draw conclusions. The module includes a number of practical examples with
  available tools, as well as an extensive demo based on analysing, visualizing and
  searching data from the music domain.
- *Module 5: Creating LOD Applications.* This module gives details on technologies
  and approaches towards exploiting LOD by building bespoke applications. In
  particular, it gives an overview of popular existing applications and introduces the
  main technologies that support implementation and development. Furthermore, it
  illustrates how data exposed through common Web APIs can be integrated with
  LOD in order to create mash-ups.
- *Module 6: Scaling up.* This module addresses the main issues of LOD and scala-
  bility. In particular, it provides gives details on approaches and technologies for
  clustering, distributing, sharing, and caching data. Furthermore, it addresses the
  means for publishing data trough could deployment and the relationship between
  Big Data and LOD, exploring how some of the solutions can be transferred in the
  context of LOD.

The skills and knowledge prerequisites associated with this curriculum are mainly
of technical nature. More specifically, in order to be able to grasp the main concepts,
the application functions and the presented approaches, some previous knowledge in IT
development and engineering are very useful. However, the lack of experience in a
particular area can be compensated for by the examples and step-by-set guides included
in the modules, which demonstrate how the learned principles and techniques can be
applied. The more advanced modules can benefit from some knowledge in the corresponding fields.

In an effort to provide high-quality training, suitable for the needs of data scientists,
the EUCLID curriculum has been through several revisions on structure, arrangement
and content after presenting it to a number of experts and gathering their feedback.

As a result of these revisions, the curriculum was refined and developed in more detail in order to include a number of expected outcome competencies, as well as a variety of exercises and examples. The content of the EUCLID modules has been redesigned to be better aligned and support a smoother process of skills built-up and development. While having an individual objective, each module contributes to further developing the skills and knowledge gained by the previous one thus aiding to acquiring an overall understanding and expertise in the field. As mentioned before, the curriculum has been constantly updated based on feedback from the community.

4 Implementing and Delivering the LOD Curriculum via OERs

In order to implement our LOD curriculum and deliver it to the data science community in the form of OERs, we have developed a production process that defines the sequence of steps for the production of OERs. Initially, 3 basic steps were planned in order to create each module and its exercises (see Fig. 3). Firstly, following the curriculum, the draft of the training material would be created, which includes slides for a webinar, as well as HTML content for online distribution. Secondly, feedback on the drafts would be gathered and analysed. Finally, based on the comments and feedback, each module would be refined before delivering an eBook encompassing all the training materials, which include written documents, examples, presentation slides, as well as the video recording of the webinar.

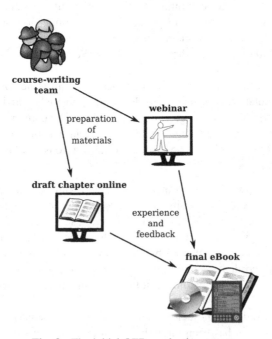

Fig. 3. The initial OER production process

During the production of the first module, this process was further refined and elaborated to include some intermediate steps (see Fig. 4). One thing that became obvious was the instrumental role of the preparation and delivery of the webinar in the production process. The webinar was therefore produced in two stages. First, an internal webinar was held in order to collect feedback from project partners about its content and structure. The learning materials were revised through collecting comments and feedback from the internal broadcasting of the first webinar and the publication of the first version of the eBook chapter.

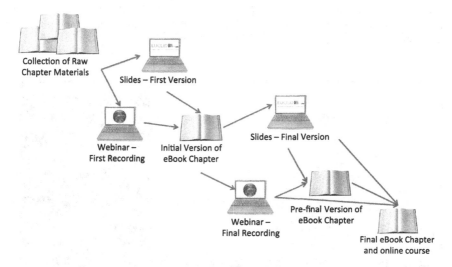

Fig. 4. The revised OER production process

Subsequently, a second version of the webinar was produced, this time publically broadcasted. Based on the community feedback received from the broadcasting of the second webinar, the structure and content of the module were finalised and the eBook chapter was produced from all the finalised content. It was also decided that additional material in the form of an online course would accompany the final eBook chapter and would be part of the training programme offered to the community. This process has been applied for the production of all EUCLID modules.

Based on the curriculum and following the OER production process, the EUCLID learning materials have been produced in various forms, in accordance with the targeted means of delivery. In particular, the EUCLID presentation slides were the first learning materials produced for each module. They provide an overview of the main concepts covered in each module and contain an extensive set of examples, so that the concepts of the module are explained to practitioners more effectively. These presentations have been published on a dedicated SlideShare channel[10].

---

[10] http://www.slideshare.net/EUCLIDproject.

Following the production of the presentation slides, a series of interactive webinars were conducted for each module. During these webinars, an expert would give a lecture on the topics covered by each module, using the module's presentation slides. The online audience had the opportunity to interact with the presenters and with each other. They were able to do so by asking questions and providing feedback about the webinar's content, either via Twitter or via a chat facility offered by the Livestream broadcasting service[11]. Figure 5 shows a still from a webinar broadcasted via Livestream. Recordings of all the webinars have been made available via the EUCLID channel on Vimeo[12]. In order to enrich our learning materials with overviews and walkthroughs of popular LOD tools and platforms, we also produced a number of screencasts. The screencasts explain in a short and effective way the tools and platforms in question.

Fig. 5. Still from the live broadcast of an interactive webinar

Additionally, a set of interactive exercises and quizzes were developed in order to enable learners to self-assess their learning progress throughout the series of modules. The quizzes consist of multiple-choice questions that test the knowledge acquired in each module (see Fig. 6). The interactive exercises allow learners to practice what they have learned in each module by using real tools, such as the Information Workbench, and real datasets, such as the MusicBrainz dataset. In these exercises, learners are offered with bespoke SPARQL endpoints, where they can try their SPARQL queries, visualise datasets, as well as develop LOD applications. We have also collected a number of exercises around LOD and semantic technologies from various summer schools. All the exercises are available in the EUCLID web site[13].

---

[11] http://new.livestream.com.

[12] http://vimeo.com/euclidproject.

[13] http://www.euclid-project.eu/resources/exercises.

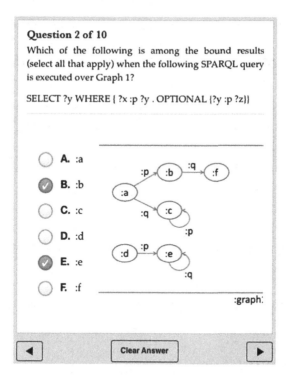

Question 2 of 10

Which of the following is among the bound results (select all that apply) when the following SPARQL query is executed over Graph 1?

SELECT ?y WHERE { ?x :p ?y . OPTIONAL {?y :p ?z}}

- A. :a
- B. :b
- C. :c
- D. :d
- E. :e
- F. :f

Clear Answer

Fig. 6. A multiple-choice question of a quiz that tests the learner's knowledge on SPARQL

Figure 7 shows the collections of the produced learning materials as these are presented in the EUCLID web site. As it can be seen, the learning materials are organised by module and format. Each module contains two types of learning materials: an eBook chapter and an online course. Both types of materials are available in multiple formats, targeting a wide range of devices.

The EUCLID eBook encompasses all the content for each module in a structured and interactive way. The eBook serves as the basis for self-learning, as well as for revisiting certain topics after a training is completed, e.g. as part of a EUCLID training event. For each module, the feedback gathered after the delivery of the presentation slides and the webinar has been used to restructure the module content for final delivery as an eBook chapter. The EUCLID eBook, therefore, represents the final outcome of the learning materials revising process.

The eBook integrates all the learning materials produced by the project. In particular, it contains multimedia and interactive elements, such as clips from the EUCLID webinars, screencasts, as well as self-assessment quizzes and exercises. The eBook is available to download from the EUCLID web site, as well as the Apple iBook Store [17].

In order to maximise the impact of the EUCLID learning materials on the community and bring them closer to as many people as possible, the EUCLID eBook has been made available for a variety of platforms and formats:

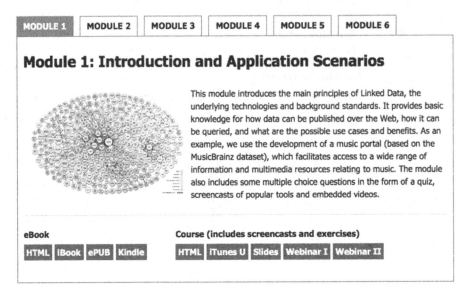

Fig. 7. The collections of learning materials offered by EUCLID, organised by module and format

- Web browsers (HTML format)
- Apple iPad and MacOS (iBook format)
- eReaders (ePUB format)
- Amazon Kindle devices (MOBI format)

The EUCLID online courses are learning pathways (or syllabi) that are based on the EUCLID learning materials. Learners can study these courses at their own pace, as there is no predetermined start or end date. The online courses have a focus on learning outcomes, which drive the organisation of the content. This means that the ultimate purpose of the online courses is to help the learner achieve the specified learning outcomes. Consequently, the EUCLID online courses differ from the EUCLID eBook in that they are shorter and targeted towards a smaller set of learning outcomes compared to the eBook, which covers a broader spectrum of LOD skills.

The EUCLID courses are available for the following platforms:

- Web browsers (HTML format)
- iTunes U on the iPad

In order to address the needs of specific data science professions, we have devised the learning pathways matrix shown in Fig. 8. These learning pathways combine the EUCLID modules towards acquiring the skills that are required by different data science professions: data architects, data managers, data analysts and data application developers, as well as different skill levels: introductory, intermediate and advanced.

	Data Architect	Data Manager	Data Analyst	Data Application Developer
Introductory Level	Module 1: Introduction and Application Scenarios	Module 1: Introduction and Application Scenarios	Module 1: Introduction and Application Scenarios	Module 1: Introduction and Application Scenarios
Intermediate Level		Module 2: Querying Linked Data	Module 2: Querying Linked Data	Module 2: Querying Linked Data
	Module 3: Providing Linked Data	Module 3: Providing Linked Data		Module 3: Providing Linked Data
			Module 4: Interaction with Linked Data	Module 4: Interaction with Linked Data
Advanced Level				Module 5: Creating Linked Data Applications
	Module 6: Scaling up	Module 6: Scaling up		

Fig. 8. The EUCLID learning pathways matrix for different data science professions and levels

5 Lessons Learned and Best Practices

Throughout the design and implementation of our curriculum, we have actively sought the input and feedback of LOD experts and the wider LOD community. In particular, we have collected feedback from community members via several LOD and Semantic Web mailing lists, as well as the EUCLID channels on Twitter and other social media. We have used the professional social network LinkedIn in order to build a dedicated EUCLID group[14] and carry out discussions with the community members. We have also collected feedback from interacting with audiences synchronously during the live broadcasting of our webinars. Additionally, we have had a number of opportunities to interact face-to-face with various audiences via training events and dedicated workshops and tutorials.

As mentioned before, the EUCLID curriculum was designed and implemented considering data practitioners and professionals as its main audience. However,

---

[14] https://www.linkedin.com/groups?gid=4917016.

the experience gathered throughout the project duration has clearly demonstrated that the produced OERs are used by a much broader audience, including researchers, students, professionals, managers, technology experts, etc. Therefore, the curriculum and the trainings can be of benefit for anyone who aims to gain a broader and deeper understanding of how to manage data in accordance with LOD principles.

We have employed two main methods for delivering our OERs: via online channels and directly by a professional trainer. Given that the main target group of the EUCLID OERs is data professionals, self-training and distance learning remain the main means of communicating the courses content. These types of training methods are relatively flexible when it comes to geographical location and time-slot allocation and are, therefore, suitable for on-the-job, but also parallel-to-the-job training.

Online communication channels, such as platforms for sharing slides, videos or complete training courses are very useful for supporting self-training and distance learning. In fact, we found out that SlideShare and Vimeo have proven to be extremely applicable in terms of sharing and disseminating the project results. The achieved outreach is much greater and really anyone interested in the topic can benefit. Furthermore, the interactive webinars have been very successful, enabling a high number of simultaneous views and overcoming geographical boundaries. The geographical distribution of the audiences that participated in the webinars is visualised in the geo-plot of Fig. 9, where one can appreciate high participation across several European countries and the United States.

In contrast to self-training based on online resources, distance-learning provides more guidance to the students in terms of the learning plan but also the support in terms of interaction with trainers or gathering feedback. It was not within the scope of EUCLID to organise a distance-learning event, however, the materials are well suited to be used as a basis for such a course. In particular, the combination of the live webinars, guided tutorials, and the official course materials can easily support such

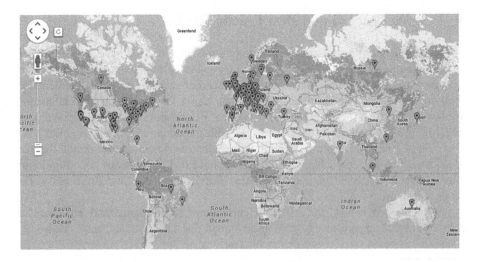

Fig. 9. Geographical distribution of the interactive webinars' live audiences

a learning approach. EUCLID's materials have also proven to be very useful as part of on-site trainings. The trainings were conducted both for professionals, as well as for people with more of a research background.

Overall, the feedback received from the community regarding the quality of the produced learning materials and the eBook has been positive. LOD experts and the broader community have appreciated the effort that was dedicated by the project partners for the production of the materials, as well as the fact that these materials cover a wide range of skills and learning objectives. Additionally, the fact that the EUCLID interactive eBook has been made available for a variety of emerging educational platforms, such as the iPad and other tablets, has also been seen as a significant advantage of the project, regarding the potential impact of the eBook on data science practitioners and therefore its sustainability and its future uptake by the community.

Throughout the duration of the EUCLID project, we have experimented with different approaches for the design and delivery of LOD OERs. The lessons we have learned have led us to finalise our curriculum and our OER production process. We have also acquired a valuable insight into the various challenges associated with the design and delivery of OERs specifically for LOD. We have thus distilled our experiences and lessons learned into a set of best practices, which is outlined in the following two sections.

5.1 Best Practices for the Design of LOD OERs

1. **Industrial Relevance**– our curriculum takes into account the needs of industry related to LOD. Future work aims to automatically mine and analyse relevant job adverts to gain desired competencies for the sector. This is supported by the following best practice.
2. **Curriculum Design Team**– where the team is composed of a number of roles to fully capture industrial, academic and pedagogical requirements. Our team comprises of industrial partners (Ontotext, FluidOps), who have extensive experience with professional training, industrial requirements and scalable tools, academic partners (KIT, STI International), who have research expertise in LOD and pedagogical experts (The Open University).
3. **External Collaborations** – to gain world-class curriculum expertise where necessary and to facilitate course delivery and dissemination.
4. **Explicit Learning Goals** – to which all learning materials (slides, webinars, eBook chapters) are developed. Learners are guided through the learning goals by learning pathways – a sequence of learning resources to achieve a learning goal.
5. **Show Realistic Solutions** – rather than mock examples we utilize systems that are deployed and used for real.
6. **Use Real Data** – we use a number of large datasets including for example, the MusicBrainz dataset that contains 100Ms of triples.
7. **Use Real Tools** – our collection of tools are used in real life, including for example Seevl, Sesame, Open Refine and GateCloud.

8. **Show Scalable Solutions** – based upon industrial-strength repositories and automatic translations, for example using the W3C standard R2RML for generating RDF from large data contained in standard databases.
9. **Eating Our Own Dog Food** – we monitor communication and engagement with the LOD community through W3C email lists, in the social network channels LinkedIn and Twitter, as well as content dissemination channels such as Vimeo and SlideShare. We transform the monitoring results into RDF and make these available at a SPARQL endpoint. In this respect we use LOD to support Learning Analytics.

5.2   Best Practices for the Delivery of LOD OERs

1. **Open to Format** – our learning materials are available in a variety of formats including: HTML, iBook (iPad and MacOS), ePUB (Android tablets), MOBI (Amazon Kindle).
2. **Addressability** – every concept in our curriculum is URI-identified so that HTML and RDF(a) machine-readable content is available.
3. **Integrated** – to ease navigation for learners the main textual content, relevant webinar clips, screencasts and interactive components are placed into one coherent space.
4. **High Quality** – we have a formalised process where all materials go through several iterations to ensure quality. For example, for each module we run both a practice and a full webinars facilitating critique and commentary.
5. **Self-Testing and Reflection** – in every module we include inline quizzes and exercises formulated against learning goals enabling students to self-monitor their progress.

6   Conclusions and Further Work

The EUCLID project has established a rigorous process for the production and delivery of OERs about LOD. This process defines a series of iterations in the production of learning materials, with multiple revisions from internal and external stakeholders, in order to ensure high quality in the produced materials. Based on our experiences and lessons learned in designing and implementing the production process, we have also established a set of best practices for the design and delivery of OERs specifically for LOD.

Building upon the lessons learned and best practices of the EUCLID project, we will be expanding our methodology in the context of the EDSA project (European Data Science Academy)[15]. Within this new project, we will produce and validate adaptable multilingual curricula, which will target the latest data science needs of industrial sectors across Europe. More specifically, we plan to establish a virtuous learning production cycle, whereby we will analyse the required sector specific skillsets for data

---

[15] http://edsa-project.eu.

analysts across the main industrial sectors in Europe; develop modular and adaptable data science curricula to meet these needs; and deliver training supported by multi-platform and multilingual learning resources based on our curricula. The curricula and learning resources will be continuously evaluated by pedagogical and data science experts during both development and deployment.

Acknowledgment. The research leading to these results has received funding from the European Community's Seventh Framework Programme under grant agreement no 296229 (EUCLID project).

References

1. Atkins, D.E., Brown, J.S., Hammond, A.L.: A review of the open educational resources (OER) movement: Achievements, Challenges, and New Opportunities. The William and Flora Hewlett Foundation (2007)
2. Attwood, R.: OER university to cut cost of degree. Times Higher Education (2011). http://www.timeshighereducation.co.uk/story.asp?storycode=415127
3. Benjamins, R., Jariego, F.: Open Data: A 'No-Brainer' for all. Telefónica Innovation Hub (2013). http://blog.digital.telefonica.com/2013/12/05/open-data-intelligence/
4. Berners-Lee, T.: Linked Data - Design Issues (2006). http://www.w3.org/DesignIssues/LinkedData.html
5. Bizer, C., Heath, T., Berners-Lee, T.: Linked data—The story so far. Int. J. Semant. Web Inf. Syst. **5**(3), 1–22 (2009)
6. Cadwalladr, C.: Do online courses spell the end for the traditional university? The Guardian (2012). http://www.theguardian.com/education/2012/nov/11/online-free-learning-end-of-university
7. Davenport, T.H., Patil, D.J.: Data Scientist: The Sexiest Job of the 21st Century. Harvard Business Review (2012). http://hbr.org/2012/10/data-scientist-the-sexiest-job-of-the-21st-century
8. Glick, B.: Government calls for more data scientists in the UK. Computer Weekly (2013). http://www.computerweekly.com/news/2240208220/Government-calls-for-more-data-scientists-in-the-UK
9. Kanwar, A., Uvalić-Trumbić, S., Butcher, N.: Basic Guide to Open Educational Resources (OER). Commonwealth of Learning & UNESCO (2011). http://www.col.org/resources/publications/Pages/detail.aspx?PID=357
10. Magoulas, R., King, J.: Data Science Salary Survey: Tools, Trends, What Pays (and What Doesn't) for Data Professionals, O'Reilly (2014)
11. Manyika, J., Chui, M., Brown, B., Bughin, J., Dobbs, R., Roxburgh, C., Byers, A.H.: Big data: The next frontier for innovation, competition, and productivity. McKinsey Global Institute (2011). http://www.mckinsey.com/insights/business_technology/big_data_the_next_frontier_for_innovation
12. McAfee, A., Brynjolfsson, E.: Big Data: The Management Revolution. Harvard Business Review (2012). http://hbr.org/2012/10/big-data-the-management-revolution/
13. McKenna, B.: Demand for big data IT workers to double by 2017, says eSkills. Computer Weekly (2012). http://www.computerweekly.com/news/2240174273/Demand-for-big-data-IT-workers-to-double-by-2017-says-eSkills

14. OEDB, 80 Open Education Resource (OER) Tools for Publishing and Development Initiatives (2007). http://oedb.org/library/features/80-oer-tools
15. Open Knowledge Foundation. Open Data Handbook Documentation (2012). http://opendatahandbook.org/pdf/OpenDataHandbook.pdf
16. Sharples, M., Mcandrew, P., Weller, M., Ferguson, F., Fitzgerald, E., Hirst, T., Mor, Y., Gaved, M., Whitelock, D.: Innovating Pedagogy Exploring new forms of teaching, learning and assessment, to guide educators and policy makers. The Open University (2012). http://www.open.ac.uk/personalpages/mike.sharples/Reports/Innovating_Pedagogy_report_July_2012.pdf
17. Simperl, E., Norton, B., Acosta, M., Maleshkova, M., Domingue, J., Mikroyannidis, A., Mulholland, P., Power, R.: Using Linked Data Effectively: iPad Edition, The Open University (2013)
18. Technology Enhanced Knowledge Research Institute (TEKRI). Open Education Resources (OER) for assessment and credit for students project: Towards a logic model and plan for action. Athabasca University (2011). http://auspace.athabascau.ca/bitstream/2149/3039/1/Report_OACS-FinalVersion.pdf
19. Wiley, D.: On the sustainability of open educational resource initiatives in higher education. OECD Centre for Educational Research and Innovation (CERI) (2007). http://www.oecd.org/dataoecd/33/9/38645447.pdf

On Some Russian Educational Projects in Open Data and Data Journalism

Irina Radchenko[1](✉) and Anna Sakoyan[2]

[1] ITMO University, St. Petersburg, Russia
iradche@gmail.com
[2] Russian Analytical Publication Polit.ru, Moscow, Russia
ansakoy@gmail.com

Abstract. The article discusses some Russian Open Educational resources, such as School of Open Data and collective blog dedicated to Data Driven Journalism. Since 2013, on the basis of this collective blog authors have been launching Data Expeditions in Russian on different topics. The most recent Data Expedition experience showed that this format can be easily integrated into traditional educational practices and actually benefits from it in terms of efficiency.

Keywords: Open data · Open education · Open educational resources · Data expedition · Data journalism

1 Introduction

The more datasets are published openly on the Internet, the more there are ways of their application. Apart from the most obvious areas, such as government, business and journalism, there have already been attempts to employ Open Data in education. One of the most prominent projects in this respect is LinkedUp initiative [1]. In fact, there are several possible directions [2], in which using Open Data in education can be regarded.

One is just to use data and the connected technologies to make educational tools more efficient and flexible. Another direction is the introduction of Open Data into the educational environment, in order to provide students with more options to organize their schedule and learning process. Last, but not least, Open Data can be viewed as the material, on which the educational process is actually built.

In this article, we will focus on the latter approach, which is how Open Data can become the object of learning and in which areas it is applicable. Our account is mainly based on our own experience in organizing data-driven educational projects with Open Data as part of the learning material.

2 Open Data School Project

One of the first open educational initiatives in Russia, which was based on the use of Open Data as learning material, was Open Data School [3]. NGO "Infoculture" [4] launched Open Data School in August 2013, under the auspices of the Open Government of Russia [5].

D. Mouromtsev and M. d'Aquin (Eds.): Open Data for Education, LNCS 9500, pp. 153–165, 2016.
DOI: 10.1007/978-3-319-30493-9_8

Open Data School classes formed a series of live lectures and workshops on the topic of Open Data and were later stored at its website as videos and their transcripts [6]. As learning material the workshops widely used Open Data stored at such resources as the Portal of the Unified Interdepartmental Information and Statistical System, Hub of Data by NGO "Infoculture" and official Open Data portals of the Russian Federation.

Classes were free for students and all the resulting materials are available under an open license.

All classes in Open Data School were divided into two main courses: The Basics of Open Data Processing and The Introduction to Data Driven Journalism. In fact, as a survey showed, 73.2 % of respondents were interested in both courses, while 21 % were keen on the Open Data direction and only 5.7 % on Data Driven Journalism alone (see Fig. 1)[1].

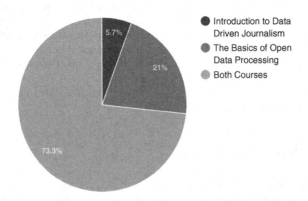

Fig. 1. Distribution of people interested in the Open Data School courses

The School courses provided guest lectures by both Russian and foreign experts, including Maxim Dubinin (NextGIS [7]), Konrad Hoeffner (University of Leipzig) and Natalia Karbasova (Hubert Burda Media). The ultimate purpose of this project was to promote Open Data among the Russian community by spreading information of how they could be used.

Over 200 students applied for the live course, male and female applicants in roughly the same proportion. The age of the participants mostly ranged from 20 to 40 years old (Fig. 2). More than 73 % of Open Data School participants had the experience of programming and /or data processing.

Open Data lectures and workshops provided basic information on how to search for datasets on the Internet and assess their quality. They also contained a profound account of major world centers of excellence for Open Data and Semantic Web, on geospatial data and geographic information systems, on open databases, open

---

[1] You can find all figures of this article in Open Data blog: http://iradche.ru/data-expeditions/data-expeditions-stats/open-data-rus-edu-projects/.

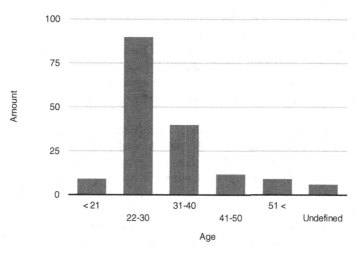

Fig. 2. Age of Open Data School participants

knowledge and Semantic Wiki, on approaches to cataloging datasets and on the practice of publishing data on the example of the pilot datasets published by the federal executive authorities.

The Data Driven Journalism track discussed the concept of journalism based on data processing, history of Data Journalism and its pioneers, online tools for data visualization, as well as the techniques of searching the required datasets on the Internet, data processing, and finally creating digital story. This included such topics as visualization of budgets, trends in the visual presentation of information all over the world, visualization of statistic datasets, construction of timelines and design of the interactive charts.

3 'Data Driven Journalism in Russia' Project

In April 2013, authors attended a seminar on Data Driven Journalism [8], which was organized by the Open Government of Russia. As a result, it was decided to launch an open Russian-language educational resource focused on Data Driven Journalism and related subjects [9]. We intended to make it a platform that accumulates corresponding materials including tutorials, collections of helpful links and instructions, both original and translated. It was supposed to be centered around a regularly updated blog and also provide pages on specific subjects.

In the beginning, the blog had only two authors. Later their number grew, as more specialists engaged with Data Driven Journalism or data analysis joined the project. By August 2015, the number of coauthors at DataDrivenJournalism.ru reached 11 people.

At the moment, the key sections at DataDrivenJournalism.ru are following.

'Practicum' section provides some information on our past data-expeditions, and also instructions and tutorials. The latter include some basic information about data and more advanced practical guidelines on how to work with spreadsheets, APIs, and visualization and analytical tools, as well as some programming languages, such as R.

The section 'Seminars' provides links to the descriptions and follow-up materials of our offline seminars on Data Driven Journalism.

'Data Journalist Tool Kit' section presents a list of links to helpful tools for data wrangling, cleaning, analysis, and visualization.

'Presentations' section is a collection of presentations, some of which are rather general and others are instructions on how to work with Github or search for data and visualize them, etc.

'Helpful Links' section is a collection of links to external sources, which we found instructive. These can be blogs, tutorials, Data Journalism projects, etc.

The home page of DataDrivenJournalism.ru shows all the recent updates, including announcements of events, tutorials (both translated and original), guest posts, reviews, walkthroughs and presentations. Some of these posts are later categorized into the sections described above.

DataDrivenJournalism.ru also has a social media presence: its updates are broadcast via its Facebook page and Twitter. It also has accounts on Pinterest and Github.

In April 2013, the founders of the blog participated in an online Data Expedition launched by School of Data [10], an educational project created by Open Knowledge Foundation (OKF, [11]). That international Data Expedition included live interaction via Google Hangout video chat and also presented a considerable number of open and free tools for both online collaboration and data processing. Another invaluable part of that expedition was an experience exchange among the participants from different countries, including Romania, Russia, South Africa, UAE, and Uganda. Generally, the team's performance was quite successful, first and foremost due to the combination of constant friendly and supportive communication among the participants. A more detailed account of that expedition can be found in a recap article by Anna Sakoyan at the School of Data's website [12].

After this Data Expedition finished, the two DataDrivenJournalism.ru authors decided to launch a Russian-language Data Expedition based on a similar model and using the experience gained within the expedition at School of Data. The key adopted principles were openness, the use of openly available tools and data and participants' cooperation as the main part of the educational process.

4 Data Expeditions

Data expeditions are educational events aimed at teaching the techniques of open data processing. Most of these events are based on mixed educational principles, which combine traditional offline teaching and online interaction.

It might be instructive to first describe this format by providing the following list of methodological recommendations, which are based on our experience of running four Data Expeditions.

1. Define the educational objectives of your Data Expedition. For instance, it can be a full-cycle process of building a digital story, where as a result of the Data Expedition participants are expected to implement a data-journalism project, starting with searching for data, through data processing and visualization and publishing

their findings in a form of a blog post or an article. Or it can be an expedition aimed at mastering some particular techniques or tools, such as searching, cleaning, visualizing, working with spreadsheets, etc.

2. Define the topic of your Data Expedition. For a full-cycle Data Expedition topics can be formulated in rather broad and general terms, such as education, budgeting, government spending, and so on. For a more narrow-scale Data Expedition, the choice of a topic rather depends on which datasets at the instructor's disposal are best suited for mastering a particular skill.

3. Create and prepare a special Google Group for the planned Data Expedition. Google Group provides a common platform for online peer-to-peer interaction among the participants. It allows them to exchange their knowledge, experience and findings, provide feedback, share their work and ask questions. This means that a group becomes an accumulator of knowledge, which is why it is important to create it as a forum and not just as a mailing list. It can also be used by instructors for publishing tasks and learning materials.

4. Create and prepare a Google Spreadsheet, which contains a list of the Data Expedition participants and their emails. It is helpful to apply different colors to a participant's row depending on their progress in task completion. It is also convenient for the assessment of the general participants' performance in the course of the Data Expedition.

The header of the spreadsheet should contain a list of tasks (their short names) that are supposed to be completed by participants. After that, the spreadsheet is ready to be shared with the participants, so that they can see their progress and submit their assignments. To protect the spreadsheet data from unintentional change by a participant, it can be shared with the commenting option only. After a task is completed, a participant is asked to add a comment with a link to the completed task to the corresponding cell. Instructors should monitor these comments and if the task is done correctly, they update the sheet by adding the link to the commented cell.

5. Use openly published learning materials as additional sources.

6. Be careful with timing. The experience of our past Data Expeditions showed that the best time for launching Data Expedition, at least in Russia, is February to April and September to November. Other months might be inconvenient for students who might be busy taking exams or leaving for holidays. However, this is rather a concern for expeditions that take place outside a curriculum. Otherwise, it can be easily transformed into an end-of-term project.

7. A Data Expedition can last from a couple of days to two weeks. Longer expeditions are possible, but seem less focused and therefore less efficient. It is important to evenly spread all the tasks during the expedition. Tasks or assignments should come after corresponding lectures or seminars, preferably on the same day. They can be both published in the Google Group or sent directly to participants via email.

It is best when there are several organizers to a Data Expedition, although we have an experience with only one-instructor (organizer) expeditions. It might be a good idea to invite some experts on the topic of Data Expedition to give a talk (either offline or online), which introduces the participants to the subject. For instance, if the topic concerns exploring some population dynamics, demographers might be quite helpful.

Next, we shall consider our general observations based on our Data Expedition experience and describe the specific of particular approaches to expeditions we employed.

As the Fig. 3 shows, the core participants of most expeditions tend to be above 21 and below 40 years old. There might be some skew towards one of the age-categories, but basically both groups are represented in a considerable proportion. Interestingly, there are normally several participants between 51 and 60 years of age. They are always few, but still they do appear in most cases (and they are definitely different people in each case). Participants between 16 and 21 years old are also rather few with the exception of fourth Data Expedition (DE4), which took place as part of a brief university course and was part of the students' curriculum.

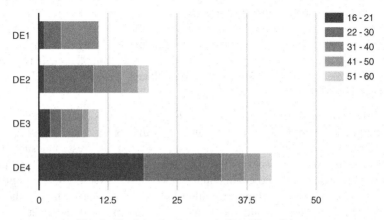

Fig. 3. Participation by age in Data Expeditions

In all the DE instances, female participants slightly outnumber male participants (Fig. 4). It might be somewhat instructive to have a closer look at the distribution of the participants' involvement based on both their gender and level of education. As the diagram shows, female participants tend to either have a complete higher education or to be undergraduate students. We have to point out here that generally the involvement of undergraduates, in the first place, is due to DE4, which was a part of a university course. As to males, we can observe a relatively higher involvement of those with 'incomplete higher education' and postgraduate students. Incomplete higher education describes a characteristic situation acknowledged by the Russian legislation when a person has successfully completed at least four terms (semesters) at university, but then discontinued their education and therefore holds no degree.

Figure 5 reflects the level of participants' basic skills by the beginning of an expedition. The key observation here is that, although all the expeditions implied that people with no experience are welcome, the number of such participants is rather low. The most common skill overall is some experience of working with spreadsheets. This is often combined with some experience of dealing with databases. Those with

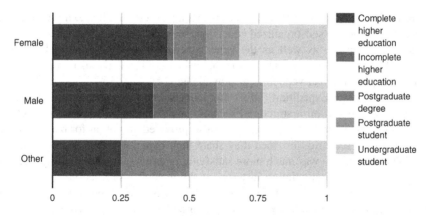

Fig. 4. Level of education by gender (through all DEs)

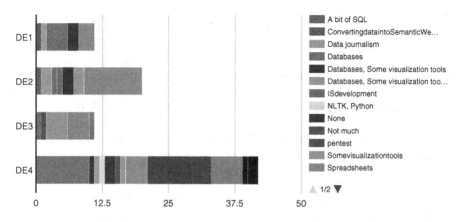

Fig. 5. Basic skills of the participants (through all DEs)

programming experience are extremely few. Judging by this tendency, we can suggest that at the moment, this kind of projects is most attractive to those with some competence in dealing with data, but not too deep competence. People with no data experience seem to be generally either not interested in acquiring this kind of skills or, which is more probable, are scared away by the prospect of dealing with something absolutely new. As to those with a higher level of expertise, they probably find this kind of activity not very exciting and unrewarding.

Now we shall provide a brief overview of each particular Data Expedition. They all represent an experimental process of building and navigating an online peer-learning process, using various approaches and testing their benefits and disadvantages.

The first Russian-language Data Expedition (DE1) took place in July 2013. Its declared objective was finding, processing and presenting data regarding universities both in Russia and around the world. DE1 did not have any particular scenario or instructions. It fully relied on participants' mutual help with some facilitation. The result was inspiring, because it demonstrated that people are interested in such activities and are willing to learn and create their data-driven projects. But it also highlighted the challenges common to self-organized education, such as the lack of instant motivation, the lack of discipline, as well as the general need for a more structured scenario. A more detailed account of DE1 was published at DataDrivenJournalism.ru (in Russian) [13]. Its English version is also available [14].

The second Data Expedition (DE2) was launched in December 2013. Unlike the first one, it was based on a structured scenario, which included an open dataset and detailed instructions on how to handle it. It also provided an option for participants to follow their own scenario, provided they share their experience and findings with the others. This experience was much more satisfying in terms of clarity and organization, although somewhat disappointing in terms of the general result, as only few participants came up with a completed final project. The reason was poor timing choice, as DE2 took place right in the face of the New Year and at the end of university semester. However, there was one full-scale project on censorship performed by a participant that followed his own scenario [15]. A detailed account of DE2 is also available both in Russian [16] and in English [17].

Particularly noteworthy is DE3 (February 2014), which was focused on rare diseases. It was more of an investigative project, rather than an educational one. It was launched in partnership with NGO "Teplitsa of Social Technologies" [18], which helped us to engage experts on rare diseases with DE3. Thus, the participants were not just working on datasets, but also had a chance dive into the specifics of the fields, which the datasets described, and enhance their understanding of their findings.

As to DE4, it was a mixed project, which combined a traditional educational approach and peer-learning practices. We shall describe this experience in a special section below.

5 Data Expeditions in Higher School of Economics and ITMO University

In 2013-2014, Irina Radchenko organized a course on "Analytical research on the Internet" at the MIEM Higher School of Economics (HSE). In that course, students were studying methods and approaches to work with Open Data. A Data Expedition was held within the course as a practical activity for the students of HSE. Students worked on the tasks both in class and at home. Unlike the abovementioned DEs, this Data Expedition was closed and focused only on the participants of the course. This is the reason why we separate it from those open online DEs.

The course mainly relied on class work, as well as work in Google Group, which registered all the important stages of work and also was the place for publishing additional teaching materials. Completion of tasks at each stage of learning was recorded in a shared Google spreadsheet and students could leave comments on each cell (which they had to accompany with their names). The instructor was the only person who had an access to editing the spreadsheet. This made the process of homework submission and assignment completion rather transparent. However, to complete the course students also had to pass a live exam. Another important task was to write an essay, which summarized the results of the accomplished work. The final grade took into account both online and offline performance.

In fact, this mixed course was to a great extent the prototype for DE4. The basic difference is that HSE's course was not open and was not aimed at building cooperative teams. The online part was rather a way to introduce new tools and practices, as well as, actually, the main source of Open Data.

6 Data Expedition in Kazakhstan

The fourth Russian-language Data Expedition (DE4) was an international project. It was held in December 2014 as part of a two week's optional practical training course on Open Data processing, which took place at KSTU (Karaganda, Kazakhstan). Namely, DE4 took place on the basis of training course for undergraduates and teachers of Karaganda State Technical University [19].

The course was built as a sequence of tasks, the implementation of which was meant lead to the creation of the final project. Each task was preceded by a lecture given by Irina Radchenko. The DE4 part, although it was basically tailored for the course purposes, was freely open for participation, so that anybody could join online. The main working platform for the online participants was a Google Group, which was both the area for peer-interaction and also provided learning materials regularly updated by the instructor, as the offline course went on. These were lecture notes, as well as presentations, and helpful links to information resources on the topic. Full lecture videos were recorded, but could not be published promptly, as this was too time-taking. The performance of the offline course participants was registered separately in a course spreadsheet, as they were eligible for certification supported by the university.

At the beginning of the course, all the participants were asked to choose a direction they are most interested in, so that they research and further work with the data on relevant topics. As a result they were grouped into several teams by interest. In each of these teams members worked on the same subject and shared their outcomes with each other, but in the end each of them was supposed to come up with their own final project.

In DE4, we used the following topics:

1. Education.
2. Culture.
3. Demographics.
4. Budget.
5. Sports.
6. Social process.

The distribution of the most popular themes is displayed in the Fig. 6.

In accordance with the structure of the training course, all working process was divided into the following stages:

– choosing a topic,
– searching for datasets on the Internet,

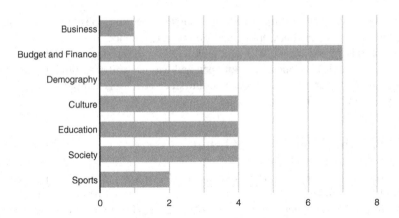

Fig. 6. Popularity of topics

- processing datasets in spreadsheets,
- building interactive infographics,
- creating a final report.

The offline course participants were required to account for each stage in a special Google Spreadsheet, where they submitted links to documents that showed the results of the work done.

Work activity within DE4 was distributed evenly and matched the stages of the offline course development. At the weekend, when there were no offline lectures, there was no significant activity in the Google Group, while during the working days (lectures were held daily) participants got involved in online communication as well.

The key working tools in DE4 were Google Documents and Google Spreadsheets, Google /Open Refine and Infogr.am. However, participants were invited to use any other tools, if they had any preferences.

All in all, 42 people registered for participation in DE4. Of these, five were from Russia, one from Ukraine, and 36 from Kazakhstan (these were the offline course participants). In fact, the actual participants of DE4 were represented only by the participants of the offline course (35 persons, or 83 % of the total number of the registered).

We shall further refer to them as 'actual participants', by which we mean those who did somehow show their presence on DE4 after the registration via the application questionnaire. We assume that some of the registered persons might have carried out tasks on their own without inform the organizers, but since it is impossible to establish, we only focus on the results available to us.

Two of the best final reports presented by the participants were published on DataDrivenJournalism.ru blog. These are a report on industry of Kazakhstan by Asylbek Mubarak [20] and a report on the Budget of Kazakhstan by Roman Ni [21].

Low activity of the participants from outside Kazakhstan can be partially explained by the fact that they did not participate in the offline course, which, as a conventional educational format, was itself a motivation to work. In addition, the offline course offered certification as an extra bonus. On the other hand, this low online activity is hardly informative as a course assessment, since the decision to open the expedition for massive participation was made right before the course began, so there was not enough time for proper online promotion, which could attract more participants.

From now on, when discussing the DE4 results, we shall count only the figures describing the performance of the actual participants who updated the performance spreadsheet. Overall, 35 participants put their names on the performance spreadsheet. However, even the first task (choosing the topic) was completed by only 25 participants (71 %).

Tasks on visualization were completed by 13 participants, and the same people submitted their final reports.

In other words, the practical part of the course was fully completed by 36 % of the actual participants and 31 % of those who initially enrolled (see Fig. 7).

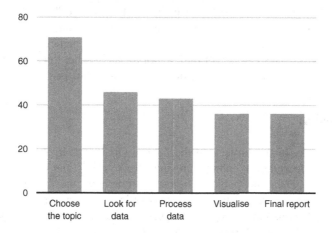

Fig. 7. Number of completed tasks at each stage (%)

Filling in the final questionnaire was not compulsory. It was filled by only 14 participants who were basically (with two exceptions) the same people as DE4 finalists. Their answers can be instructive for further methodological improvement, especially in terms of allocation of time and interaction with other participants.

To the question whether their participation in DE4 was useful, all the respondents answered affirmatively. Some pointed out that thanks to DE4 they acquired new skills and found out about interesting online tools and services for working with datasets.

Among the answers positive feedback prevails. First of all, participants liked the instructor's work and the options to do their practical tasks online. Many people mentioned the friendly atmosphere of mutual support. One respondent, however, noted that the lack of competition made their experience less exciting than it could be. Almost all the respondents expressed their desire to participate in further Data Expeditions.

Compared to the previous three DEs, DE4 seems to be a considerable progress. First of all, it is due to the relatively high percentage of the completed final projects (30 % of the total number of the registered participants). Another improvement is even distribution of the participants' efforts during the whole period of the course.

It might be tempting to draw a conclusion that the key factor here was that the DE4 was not an independent online project, but was just a part of a traditional offline course.

However, DE4 had a number of other characteristics, which could also have an impact, and these characteristics should be taken into account as a prompt on how to make independent online expeditions more motivating and efficient.

DE4 had a single central scenario and an instructor. Although experience exchange and peer-learning were strongly encouraged, there was an instructor who helped to formulate priorities, provided learning materials and assignments and assessed the performance.

As a result of successful implementation of tasks, participants could get certificates. In the case of DE4, these were KSTU certificates, which in itself may be a good motivation, because they are formal and can be referred to in CVs. An informal course cannot offer this kind of certification, but the international experience shows that the role of informal markers of success, such as so-called "badges", can be a good substitute, even if they are only relevant within the informal community inside the course.

An interesting experience was gradual assessment of the participants' work in each phase of the course. This can be not only a tool for the teacher to control the learning process, but also a motivating indicator of the progress for the participants themselves.

7 Conclusion

There are few educational projects in the field of Open Data in Russia. At the same time, there are opportunities, enthusiasm, and need to create and launch new training courses on Open Data and public information. For now, the existing courses and educational initiatives are mostly aimed at covering the very basics and are rather general. It might be also interesting to organize more specialized educational projects, for instance, on Urban Informatics, advanced Data Driven Journalism, Statistical studies, and so on. All such initiatives would extremely benefit from using Open Data as learning material.

Data Expeditions proved to be a helpful format, when it comes to educating people on how to use Open Data and which online services are available for searching, processing and visualizing data. It is basically a project-oriented format, which makes it possible to immediately apply newly learnt skills to a specific task with a prospect of getting an actual result. Another important aspect of this format is team work based on peer-learning approach and experience exchange. Expert consultations on Data Expedition topic, as well as encouraging participation by introducing some stimuli (including symbolical badges) can significantly contribute to the efficiency of the learning process. Last, but not least, the mixed format, which combines offline and online activities, tends to result in better performance.

The implementation of Open Education approach can significantly expand the audience of participants in training courses, as well as create an enable a productive environment for the emergence of new creative ideas on the use of the open methodology, Open Data and the methods of processing and presenting Open Data. In this respect, such informal peer-learning projects as Data Expeditions and, broader, connectivity massive open online courses (cMOOCs) are a very promising direction, because of their low cost, broad availability and an easily reproduced flexible model, which makes them both an excelllent tool of educational self-organization, as well as a very efficient supplement to a traditional training course.

References

1. LinkedUp - Linking Web Data for Education Project. http://linkedup-project.eu/
2. LinkedUp: Linking Open Data for Education. http://www.ariadne.ac.uk/issue72/guy-et-al
3. Open Data School. http://opendataschool.ru/
4. NGO Information Culture. http://www.infoculture.ru/
5. Open Government of Russia. http://open.gov.ru/
6. Training Materials of Open Data School. http://opendataschool.ru/tag/materials/
7. NextGIS. http://nextgis.ru/en/
8. Data Journalism Has Come to Russia. A Recap of the First Open Government Seminar (in Russian). http://open.gov.ru/events/5508/
9. A Little Backround Story (in Russian). http://www.datadrivenjournalism.ru/2013/04/prolog/
10. Data Expeditions. http://schoolofdata.org/data-expeditions/
11. Open Knowledge Foundation. https://okfn.org/
12. Data MOOC: Results, Findings and Recommendations. http://schoolofdata.org/2013/06/25/data-mooc-results-findings-and-recommendations/
13. Account of the First Russian-language Data Expedition (in Russian). http://www.datadrivenjournalism.ru/2013/08/first-russian-data-expedition-report/
14. First Data Expedition in Russian: Mission Complete. https://ourchiefweapons.wordpress.com/2013/08/05/first-data-expedition-in-russian-mission-complete/
15. Happy new year to strengthen censorship! (in Russian). http://echo.msk.ru/blog/pirate_jack/1231160-echo/
16. Report on the Second Data Expedition in Russian (in Russian). http://www.datadrivenjournalism.ru/2014/01/de2-report/
17. Second Data Expedition in Russian: Mission Accomplished. https://ourchiefweapons.wordpress.com/2014/01/04/second-data-expedition-in-russian-mission-accomplished/
18. NGO: Teplitsa of Social Technologies. https://te-st.ru/
19. Karaganda State Technical University. http://www.kstu.kz/
20. Asylbek Mubarak: Industry in Kazakhstan (DE4) (in Russian). http://www.datadrivenjournalism.ru/2015/01/de4_project_industry_kz/
21. Roman Ni: Budget of Kazakhstan (in Russian). http://www.datadrivenjournalism.ru/2015/01/de4_project_budget_kz/

The Open Education Working Group: Bringing People, Projects and Data Together

Marieke Guy[✉]

The Quality Assurance Agency for Higher Education, Gloucester, UK
m.guy@qaa.ac.uk

Abstract. New areas of research and discovery benefit from a space in which a community can discuss key questions, interact and be allowed to grow. For the open education data community these questions are still crude and fundamental. For example: what exactly is open education data, why is open data relevant to education and how can we use open datasets to meet educational needs? The Open Education Working Group inaugurated by Open Knowledge, a community-based not-for-profit organisation that promotes open knowledge in the digital age, is beginning to explore the relationship between data and education by exposing projects and bringing together people working on related activities. This chapter will consider initial discussions and explorations related to open data in education that have taken place through recent online activity and workshops. It will also look at the creation of a working group established with the intention of taking these discussions further.

Keywords: Open data · Open education · Community

1 Introduction

Open data in education is a relatively new area of interest with only dispersed pockets of exploration having taken place worldwide, these initial explorations will be covered later in this chapter. The phrase 'open educational data' remains loosely defined but might be used to refer to:

- all openly available data that could be used for educational purpose
- open data that is released by education institutions

Understood in the former sense, open educational data can be considered a subset of open educational resources (OERs) where datasets are made available for use in teaching and learning. These datasets might not be designed for use in education, but can be repurposed and used freely.

In the latter sense, the interest is primarily around the release of data from academic institutions about their performance and that of their students. This could include:

- Reference data such as the location of academic institutions
- Internal data such as staff names, resources available, personnel data, identity data, budgets

© Springer International Publishing Switzerland 2016
D. Mouromtsev and M. d'Aquin (Eds.): Open Data for Education, LNCS 9500, pp. 166–187, 2016.
DOI: 10.1007/978-3-319-30493-9_9

- Course data, curriculum data, learning objectives,
- User-generated data such as learning analytics, assessments, performance data, job placements
- Benchmarked open data in education that is released across institutions and can lead to change in public policy through transparency and raising awareness.

The World Economic Forum report on education and skills [5] sees there as being two types of education data: traditional and new. Traditional dataset include identity data and system-wide data, such as attendance information new datasets are those created as a result of user interaction, which may include web site statistics, and inferred content created by mining datasets using questions.

Whatever the classification it is clear that open education datasets are of interest to a wide variety of people including educators, learners, institutions, government, parents and the wider public. Some will have a passion for improving teaching and learning or have a vested interest in a particular individual's education, while others will be mining data to influence policy decisions or exploring monetisation of datasets. Open education data holds huge potential for many and its exploration is both inevitable and necessary.

2   Establishment of an Open Education Working Group

One of the goals of the LinkedUp Project [3][1], which focused on the exploitation and adoption of public, open data available on the Web by educational organisations, was to grow a community of linked data and open data practitioners that would continue to network with one another after the end of the project lifecycle. The expectation was that this community would continue to build applications, educate each other and others, and influence the future of linked and open data in an educational context.

In an effort to see discussions around open data in education pulled into the wider debate around open education, the LinkedUp Project dissemination partner, Open Knowledge[2], established the Open Education Working Group.[3] Open Knowledge is a not-for-profit organisation that promotes open knowledge, including open content and open data. It provides open data services and is the creator and licensor of CKAN[4], the worlds leading software for open data portals. Open Knowledge has been working in the open data space since 2004 and are world experts around open data, open content, principles, standards and practice in open. In 2006 Open Knowledge began work on the Open Definition[5] which sets out principles that define "openness" in relation to data and content. The definition makes precise the meaning of "open" in the terms "open data"

---

[1] http://linkedupproject.eu/.
[2] https://okfn.org/.
[3] http://education.okfn.org/.
[4] http://ckan.org/.
[5] http://opendefinition.org/.

and "open content" and thereby ensures quality and encourages compatibility between different pools of open material. Its development and use has been key in the open movement. Open Knowledge and the Open Definition Advisory Council recently announced the release of version 2.0 of the Open Definition. The short version of the definition is given as: "Open data and content can be freely used, modified, and shared by anyone for any purpose". The Definition "sets out principles that define openness in relation to data and content" and plays a key role in supporting the growing open data ecosystem.

Open Knowledge co-ordinates over 20 domain-specific Working Groups that focus on discussion and activity around a given area of open knowledge. The Open Education Working Group joins others looking at areas including open access, open science, open economics, open spending and open government data. The Open Education Working Groups goal is to initiate global cross-sector and cross-domain action that encompasses the various facets of open education, including open data. It brings together people and groups interested in the various facets of open education, from OERs and changing teaching practices, to licensing and emerging open policy. However it differs from other online discussion groups that deliberate on open education topics. This is because its goal is to seed and support trans-global, cross-sector and cross-domain activities and projects. The group also provides an opportunity for collaboration across organisations through engagement with existing groups.

3   Group Launch

The Open Education Working Group officially launched to a physical audience of over 100 in September 2013 at OKConference[6] in Geneva at a LinkedUp Open Education Panel Session titled, The facets of open education (See Fig. 1).

The panel session examined the different 'faces' of open education. The premise was that while many other open education groups exist their focus tends to be national and subject specific, considering one particular area of open education. This panel session, and in conclusion the working group itself, was an attempt to explore the synergies between different areas of open education. The panel was moderated by Doug Belshaw, Badges and Skills Lead, Mozilla Foundation. The panelists were:

- Jackie Carter, Senior Manager, MIMAS, Centre of Excellence, University of Manchester, who gave the Open Educational Resources perspective
- Davide Storti, Programme Specialist, Communication and Information Sector (CI), United Nations Educational, Scientific and Cultural Organization (UNESCO) who gave us the open practitioner perspective.
- Mathieu dAquin, Research Fellow, Knowledge Media Institute, Open University, UK, who is a LinkedUp Project team member gave the open data perspective.

---

[6] http://okcon.org/.

Fig. 1. OKCon Open Education Panel session. From left to right: Mathieu dAquin, Davide Storti, Jackie Carter and Doug Belshaw.

A write up of the launch is available from the LinkedUp Project blog[7]. Slides from the session are available on Slideshare.[8]

4   Group Structure

The Open Education Working Group has taken the approach of building its governance and defining its member structure in consultation with the wider community. This happens through bi-monthly working group calls[9] which are open to all and through use of a charter which sets out the member structure[10]. The group also participated in an open consultation process and appointed an Advisory Board which contains high-profile Open Education advocates who are experts in the field[11]. The Advisory Board provides thought leadership about the direction of the working group and helps to raise the profile of the working group by talking about the group and its work at conferences and events.

The current Advisory Board has 6 members:

- Karien Bezuidenhout, Chief Operating Officer at the Shuttleworth Foundation
- Lorna M. Campbell, Assistant Director of the Centre for Education, Technology and Interoperability Standards
- Dr. Cable Green, Director of Global Learning at Creative Commons
- Joonas Mkinen, Finnish maths teacher carrying out exciting open text book work

---

[7] See http://linkedup-project.eu/2013/09/17/open-education-panel-session/.
[8] See http://www.slideshare.net/MariekeGuy/linkedup-open-education-panel-session.
[9] See http://education.okfn.org/working-group-calls/.
[10] http://education.okfn.org/charter/.
[11] See http://education.okfn.org/advisory-board/.

– Bernard Nkuyubwatsi, initiator of the Open Education Rwanda Network
– Rayna Stamboliyska, founder of RS Strategy and OpenMENA

The Open Education Working Group website includes a blog, information on how to get involved and details of the mailing list and activities. The blog features regular guest blog posts including a series of Open Education Around the World blog posts currently covering 15 countries from Europe, Asia, Africa and America[12]. There is also an online interactive open education timeline which capture events, projects and activities related to open education all around the globe. The group has an active mailing list[13] and a Twitter feed[14].

The Open Education Working Group aims to be more than just a place for discussion, current actions include support for LRMI initiatives[15], standards, supporting a platform for open standards work, promotion of multilingualism for OERs, support for member activities, connections with local Open Knowledge groups (Belgium, Finland, Brazil) etc.

5   Open Education Working Group Events

Concrete activities initiated by the group have so far comprised of online discussions relating to the definition of open data in education, community webinars highlighting research and real-world activities and the collaborative writing of a living web document entitled the Open Education Handbook. We explore these activities in more detail below.

5.1   Open Education Smörgåsbord

The Open Education Working Group delivered its first official group workshop at the Open Knowledge Festival in July 2014 in Berlin[16]. The Open Education Smörgåsbord session[17] featured 6 different tables of activity, from open badges and open data to OERs for teachers and Open education policy, provided by 8 Open Education Working Group members. The tables were facilitated by Kristina Anderson from Creative Commons Sweden Miska Knapek, an information experience designer from Denmark involved with Open Knowledge Finland Irina Radchenko, Associate Professor at the Higher School of Economics in Moscow Tom Salmon, a teacher and open development researcher Darya Tarasowa Darya, maintainer of SlideWiki.org Alek Tarkowski, director of Centrum Cyfrowe, Polish NGO focusing on open issues and Public Lead of Creative Commons Poland and European Policy Fellow with Creative Commons and Marieke Guy, the Open Education Working Group Co-ordinator. One of the activities was Tom Salmons wall of open data case studies (See Fig. 2).

---

[12] See http://education.okfn.org/world/.
[13] http://education.okfn.org/mailing-list/.
[14] http://twitter.com/okfnedu.
[15] Learning Resource Metadata Initiatives - http://www.lrmi.net/.
[16] See http://2014.okfestival.org/.
[17] See http://linkedupproject.eu/2014/07/22/okfestivaling-for-linkedup/.

Fig. 2. Wall of open data in education case studies (CC BY) by Gregor Fischer (http://www.gfphotography.de/).

These covered a variety of different areas:

- AREA 1 USING DATA IN EDUCATION: Looking at use of open data to enrich teaching and learning, and to learn about data and data analysis within education or lessons.
- AREA 2 METADATA and OPEN EDUCATION RESOURCES: Looking at how metadata and open data can support education, for example through the use of OERs in universities and schools and the role of initiatives such as the learning resource metadata initiative (LRMI).
- AREA 3 GOVERNMENT DATA for EDUCATION: Looking at ways that different governments are making contributions with open data to improve education in Brazil, Holland, New Zealand and the UKl.
- AREA 4 LEVERAGING OPEN BADGES IN EDUCATION: Looking and learning about how open badges (which leverage different kinds of metadata) can be used to support and extend formal education, and personalise it in different ways.
- AREA 5 MOBILE LEARNING with OPEN DATA: Looking at how free, open source (FOSS) app authoring tools can be used to build apps that use open datasets with all kinds of applications.

In area 1 particular attention was paid to previous LinkedUp Competition entries, for example there was an explortation of how we can use open data to learn about data and data analysis within education or lessons[18] or with specific focus issues (e.g. sustainability with Globetown[19] or politics with Polimedia[20]).

5.2 'Making It Matter' Workshop

The Making it Matter workshop[21] which used the subtitle 'Supporting education in the developing world through open and linked data', was held in central London on 16th May 2014. The workshop was organised by the LinkedUp Project and the Open Education Working Group in collaboration with associate partner The Commonwealth of Learning[22].

The workshop had two real aims: firstly, it focused discussion around real-world requirements in the developing world that could be aided through the releasing of data and/or the building of relevant applications and prototypes. Secondly, it explored the opportunity to look at tools developed through the LinkedUp Challenge (the Veni and Vidi Competitions) and See how they could be used in the developing world.

The workshop was attended by approximately 30 delegates: teachers, educators, members of the open development movement, open data and linked data communities, developers and technologists. It combined talks, tool demonstrations and break-out group sessions. As the event focused on education in the developing countries, it was felt to be very important to make provisions for those who could not physically attend. Video for the entire day, including break out sessions, was streamed and shared afterwards. All breakout activities were carried out in online etherpads and a form was used to facilitate questions from remote participants.

Concrete outputs from the day were formulated around the answers to three questions:

– What real world problems are there related to education in the developing world that could potentially be solved by data and technology solutions?
– What data is out there and what data could be released to aid education in the developing world?
– Next Steps what are we going to do?

These discussion outputs fed into the requirements for a focused track for the LinkedUp Vici Competition that looked for educational applications that target developing countries.

5.3 Online Community Session

In May 2014 LinkedUp and the Open Education Working Group took part in an Open Knowledge Community Session entitled 'What has open data got to

---

[18] TUVALABS https://tuvalabs.com/.
[19] http://www.globetown.org/.
[20] http://www.polimedia.nl/.
[21] See http://linkedup-project.eu/making-it-matter-workshop/.
[22] http://col.org.

do with education'? The session was facilitated by Heather Leson (community builder at Open Knowledge) with talks from Marieke Guy (Open Education Working Group, LinkedUp Project) and Otavio Ritter (Open education data researcher, Getulio Vargas Foundation, Brazil). Otavio shared his findings from work on paper involving a comparative analysis of school open data in England and Brazil and the availability (transparency) of government information related to primary/secondary education area. Video and slides are available online.[23]

6   Open Data in Education Discussions

Through events and the mailing list the working group has supported dialogue and discussion around open data in education. These discussion topics have been explored using blog posts and in the Open Education Handbook. The main areas of dialogue have been: consideration of drivers for open data support for finding open data and analysis of use cases for open data. Elaboration on these discussion areas is given below.

6.1   Drivers for Open Data in Education

The current main drivers for open data use in education are principle, policy and practice. The charitable mission of education can be helped through a commitment to open data, enabling educators and institutions to engage with learners more effectively and in better ways. Data openness and exchange can drive quality research (collaboration, testing, replication) while promoting the social role and place of institutions themselves, helping maintain public and political commitment to the institution and making it more transparent. Education institutions are already subject to freedom of information, but new open research data policies (such as the UKs Higher Education Funding Council for England (HEFCE) consultation on inclusion as part of next Research Excellence Framework[24]) may alter obligations. In the UK, for example, large amounts of institutional data (finance, student performance, etc.) are already collected by the Higher Education Statistics Agency (HESA) and the Universities and Colleges Admissions Service (UCAS) and made widely available, and this is a trend which can be observed in many countries. The next logical step is for more open data about institutions to be made available. With agreed frameworks and metrics in place it will be easier to substantiate comparisons and claims about widening participation, or student performance, for example.

Institutions can use their own data to inform decisions and management practices, and improve business and pedagogical intelligence. By linking across other open datasets and curating the most relevant information staff and students can be supported in teaching and learning.

---

[23] See http://linkedupproject.eu/2014/07/01/community-session-what-has-open-data-got-to-do-witheducation/.

[24] http://www.hefce.ac.uk/whatwedo/rsrch/rinfrastruct/oa/.

6.2 Types of Open Data in Education

There are many different types of data that can be relevant to education and come from education. Relevant sources might include:

- Publications and literature: ACM, PubMed, DBLP (L3S), OpenLibrary, etc.
- Domain-specific knowledge and resources: Bioportal for Life Sciences, etc.
- Historic artefacts in Europeana, Geonames, etc.
- Cross-domain knowledge: DBpedia, Freebase, etc.
- (Social) media resource metadata: BBC, Flickr, etc.

Explicitly educational datasets and schemas include:

- University Linked Data: e.g. The Open University UK [1][25]
- Southampton University, University of Munster (DE), education.data.gov.uk, etc.
- OER Linked Data: mEducator Linked ER[26], OpenLearn, etc.
- Schemas: Learning Resource Metadata Initiative (LRMI[27], mEducator Educational Resources schema[28]
- Learning Analytics and Knowledge (LAK) Dataset [2][29]
- Vast Open Educational Resource (OER) and MOOC metadata collections (e.g. OpenCourseware, OpenLearn, Merlot, ARIADNE)
- UK Key Information Set[30]
- Education GPS is the OECD source for internationally[31], as well as comparable data on education policies and practices, opportunities and outcomes. Accessible any time, in real time, the Education GPS provides you with the latest information on how countries are working to develop high-quality and equitable education systems.

There are also many different ways to categorise this data.

- Student data: attendance, grades, skills, exams, homework, etc.
- Course data: employability related to courses, curriculum, syllabus, VLE data, number of textbooks, skills, digital literacy, etc.
- Institution data: location data, success/failure rates, results, infrastructure, power consumption, location, student enrolment, textbook budget, teacher names and contracts, drop out rates, total cost of ownership, sponsorship, cost per pupil, graduation rates, male vs female, years in education, ratio of students to teaching staff, etc.

---

[25] http://data.open.ac.uk.
[26] http://ckan.net/package/meducator.
[27] http://www.lrmi.net.
[28] http://purl.org/meducator/ns.
[29] http://solaresearch.org/initiatives/dataset/.
[30] http://unistats.direct.gov.uk/.
[31] http://gpseducation.oecd.org/.

- User-generated data: learning analytics, assessments, performance data, job placements, laptop data, time on tasks, use of different programmes/apps, web site data, etc.
- Policy/Government data: equity, budgets, spending, UNESCO literacy data, deprivation and marginalisation in education, participation, etc.

Other approaches to categorisation have been suggested. Louis Coiffait, then at Pearson, offered the following categories in an exploratory presentation given at the Education Innovation conference in Manchester: type (similar to listed above), purpose (intentional use), level (regional, national, international etc.)[32]. Pearson Blue Skies were responsible for a report entitled "How Open Data, data literacy and Linked Data will revolutionise higher education"[33] written in 2011. Octavio Ritter, Open education data researcher, Getulio Vargas Foundation, Rio de Janeiro, Brazil, offered other categories: macro (education policy), meso (school management) and micro (student level).

In addition to information about open licensing, a more detailed description of an open dataset may include:

- Provenance
  - Reference (government data, geo-data, etc.) – e.g. national curriculum data
    * Location of schools, universities, etc.
  - Core/Internal (course catalogue, course resources, staff data, buildings, etc.)
  - User-generated/contributed (user activities, assessments, etc.)
- Granularity
  - individual/personal
  - aggregated/analytics
  - report
- Descriptiveness
  - data streams (multimedia resources)
  - data content (textual content, database)
  - resource metadata
  - content metadata
  - paradata (as in metadata about data collection)
- Content
  - Usage/activity data (paradata as in the learning analytics definition)
  - student personal information
  - student profiles (interest, demographics, etc.)
  - student trajectories
  - curriculum / learning objectives / learning outcomes
  - educational resources (multimedia or not)
  - resources metadata (including library collections, reading lists, Talis Aspire)

---

[32] See http://www.slideshare.net/louiscrusoe/open-education-data.

[33] http://pearsonblueskies.com/2011/how-open-data-data-literacy-and-linked-data-will-revolutionise-higher-education/.

- assessment/grades
- institutional performance (e.g., ofsted, Key Information Set)
- resource outputs (publication repositories, etc.), research management data (projects and funding, etc.), research data
- cost and student funding data, budgets and finances
- classifications/disciplines/topics (e.g. JACS)

6.3   Finding Open Data in Education

One good source of open data is governments, who increasingly make data about their citizens available online. Examples from the UK include school performance data[34], data on the location of educational establishments[35] and pupil absenteeism[36]. There is also data from individual institutions such as that collated on linked universities[37] and on data.ac.uk[38] and from research into education, such as the Open Public Services Network report into Empowering Parents, Improving Accountability.[39]

Previously much of the release and use of open educational datasets has been driven by the need for accountability and transparency. A well-cited global example has been the situation in Uganda where the Ugandan government allocated funding for schools, but corruption at various levels meant much of the money never reached its intended destination. Between 1995 and 2001, the proportion of funding allocated which actually reached the schools rose from 24 % to 82 %. In the interim, they initiated a programme of openly publishing data on how much was allocated to each school. There were other factors but Reinikke and Svenssons analysis [4] showed that data publication played a significant part in the funding increase.

However recent developments, such as the current upsurge of open data challenges (see the ODI Education: Open Data Challenge[40] the LAK data challenge[41] and Open Education Challenge[42], an EU funded initiative to support projects who receive mentoring and seed funding through the European Incubator for Innovation in Education, have meant that there is an increasing innovation in data use, and opportunities for efficiency and improvements to education more generally. Their potential use is broad. Datasets can support students through creation of tools that enable new ways to analyse and access data, for example maps of disabled access and by enriching resources, making it easier to share

---

[34] See http://www.education.gov.uk/schools/performance/download_data.html.
[35] See http://data.gov.uk/dataset/location_of_educational_establishments.
[36] See http://data.gov.uk/dataset/pupil_absence_in_schools_in_england.
[37] See http://linkeduniversities.org/lu/index.php/datasets-and-endpoints/.
[38] See http://www.data.ac.uk/data.
[39] See http://www.thersa.org/action-research-centre/community-and-public-services/
2020-public-services/open-public-services-network/empowering-parents,
-improving-accountability.
[40] http://theodi.org/education-open-data-challenge-series.
[41] http://www.solaresearch.org/events/lak/lak-data-challenge/.
[42] http://openeducationchallenge.eu.

and find them, and personalize the way they are presented. Open data can also support those who need to make informed choices on education, for example by comparing scores, and support schools and institutions by enabling efficiencies in practice, for example library data can help support book purchasing.

As part of the Open Data Challenge Education, the Open Data Institute has compiled a set of interesting resources[43], including a list of potentially interesting datasets[44].

Education technology providers are also starting to see the potential of data-mining and app development. For example open education data is a high priority area for Pearson Think tank[45]. Back in 2011 they published their blue skies paper "How Open Data, data literacy and Linked Data will revolutionise higher education"[46]. Ideas around how money, or savings, can be made from these datasets are slowly starting to surface.

6.4   Using Open Data in Education

Schools Sector. Much of the innovative activity around open education data use has focused on the schools sector. Tools highlighted in the aforementioned Open Data Challenge include Locrating[47], defined as 'to locate by rating: they locrated the school using locrating.com' combines data on schools, area and commuting times Schools Atlas[48], creates an interactive online map providing a comprehensive picture of London schools, current patterns of attendance and potential future demand for school places. Data behind the atlas is available from the data store. RM Schoolfinder[49] which allows you to compare and contrast different schools, find out about what they excel at and how well children do academically. Most of the information comes from official statistical releases published by the Department for Education and Ofsted including School Performance Tables, GCSE Subject Results, school information from Edubase and summaries of the Ofsted school inspection report. Guardian GCSE schools guide[50] designed to help parents find and research local schools in England. Search by postcode to find which schools offer individual subjects, and compare how they have performed in GCSE results. Data is supplied by the Department of Education. School impact measures are based upon FFT contextual value-added scores by permission of FFT Education Ltd. Ofstead School

---

[43] See https://hackpad.com/Education-Open-Data-Challenge-kLW3ZeR98lj.

[44] See https://docs.google.com/a/okfn.org/spreadsheet/ccc?key=0Aswdg5Zc6wBhdD JsODRYMl9OS1BWY3pwYjVNR2JtSnc&usp=drive_web.

[45] See http://thepearsonthinktank.com/research/education-data/.

[46] http://pearsonblueskies.com/2011/how-open-data-data-literacy-and-linked-data-will-revolutionise-higher-education/.

[47] http://www.locrating.com/.

[48] http://www.london.gov.uk/priorities/young-people/education-and-training/london-schools-atlas.

[49] http://home.rm.com/schoolfinder/.

[50] http://www.theguardian.com/education/gcse-schools-guide.

Data Dashboard[51] provides a snapshot of school performance at Key Stages 1,2 and 4. The dashboard can be used by school governors and by members of the public to check the performance of the school in which they are interested. The data is available in RAISEonline – you will need to login to access the data and not all is openly available.

The UK is not alone in seeing the benefit of open education data, in Holland, for example, the education department of the city of Amsterdam commissioned an app challenge similar to the current ODI one mentioned earlier. The goal of the challenge was to provide parents with tools that help them to make well-informed choices about their children. A variety of tools were built, such as schooltip.net, 10000scholen.nl, scholenvinden.nl, and scholenkeuze.nl. The various apps have now been displayed on an education portal focused on finding the 'right school'. RomaScuola[52], developed under the umbrella of the Italian Open Data Initiative, allows visitors to obtain valuable information about all schools in the Rome region. Distinguishing it from some of the previous examples is the ability to compare schools depending on such facets as frequency of teacher absence, internet connectivity, use of IT equipment for teaching, frequency of students' transfer to other schools and quality of education in accordance with the percentage of issued diplomas.

Also in Europe E-school Estonia[53] provides an easy way for education stakeholders to collaborate and organize teaching/learning information. The system has a range of different functions for its various users. Teachers enter grades and attendance information in the system, post homework assignments, and evaluate students' behavior. Parents use it to stay closely involved in their children's education. With the help of round-the-clock access via the internet, they can see their children's homework assignments, grades, attendance information and teachers notes, as well as communicate directly with teachers via the system. Students can read their own grades and keep track of what homework has been assigned each day. They also have an option to save their best work in their own, personal e-portfolios. District administrators have access the latest statistical reports on demand, making it easy to consolidate data across the district's schools.

Another interesting project is Social Accountability for the Education Reform in Moldova, a website for enabling the public to monitor the schools performance[54]. The site includes the planned expenditures for all the schools in Moldova (2014)[55]. The School Portal[56], developed under the Moldova Open Data Initiative[57], uses data made public by the Ministry of Education of Moldova to offer comprehensive information about 1529 educational institutions in the

---

[51] http://dashboard.ofsted.gov.uk.

[52] http://lab.evodevo.it/romascuola/viewer.

[53] http://e-estonia.com/component/e-school/.

[54] http://expert-grup.org/en/proiecte/item/916-gpsa-moldova.

[55] http://www.budgetstories.md/bugetul-scolii-2014/.

[56] http://afla.md/.

[57] http://data.gov.md/en/.

Republic of Moldova. Users of the portal can access information about schools yearly budgets, budget implementation, expenditures, school rating, students' grades, schools' infrastructure and communications. The School Portal has a tool which allows visitors to compare schools based on different criteria infrastructure, students' performance or annual budgets. The additional value of the portal is the fact that it serves as a platform for private sector entities which sell school supplies to advertise their products. The School Portal also allows parents to virtually interact with the Ministry of Education of Moldova or with a psychologist in case they need additional information or have concerns regarding the education of their children.

Further afield in Tanzania Shule.info allows comparison of exam results across different regions of Tanzania and for users to follow trends over time, or to see the effect of the adjustments made to yearly exam results. The site was developed by young Tanzanian developers who approached Twaweza, an Open Development Consultant, for advice, rather than for funding. The result is beneficial to anyone interested in education in Tanzania. In Kenya the Open Institute used data collected from the Kenya National Examinations Council (KNEC) and the Kenya Open Data Portal to release KCPE Trends[58] a simple tool designed to visualise Kenya Certificate of Primary Education (KCPE) performance records of primary schools in Kenya from 2006 to 2011.

In Burkina Faso they have opened their open data portal[59]. The open data team of the government have worked with civil society and some start-up to realise a pilot project that consist on visualizing on a map the primary schools of a municipality. In addition, some important indicators for Burkina were present. Those indicators (proximity of canteen, latrine, or potable water point) can help parents choose the best school for their children, investors to choose the better place to build a school, or the government itself to measure the impact of its actions. They also have information on success rates in examinations, the number of classes, the number of teachers, the number of girls and boys, the geo-localisation of the school, and also display a picture of the school. In Brazil the school census collects data about violence in school area (like drug traffic or other risks to pupils). Based on an open data platform developed to navigate through the census, it was possible to see that, in a specific Brazilian state, 35 % of public schools had drug traffic near the schools. This fact created a pressure in the local government to create a public policy and a campaign to prevent drug use among students[60]. Further information is provided in Open Data for Education in Brazil[61].

Similar activity is happening in North America, Canada, Australia and New Zealand. Discover Your School, developed under the Province of British

---

[58] http://apps.openinstitute.com/kcpetrends.

[59] data.gov.bf.

[60] See https://www.facebook.com/media/set/?set=a.484468108297027.1073741826.
273872446023262&type=3.

[61] http://stop.zona-m.net/2013/03/open-data-for-education-in-brazil/.

Columbia of Canada Open Data Initiative[62], is a platform for parents who are interested in finding a school for their kids, learning about the school districts or comparing schools in the same area. The application provides comprehensive information, such as the number of students enrolled in schools each year, class sizes, teaching language, disaster readiness, results of skills assessment, and student and parent satisfaction. Information and data can be viewed in interactive formats, including maps. On top of that, Discover Your School engages parents in policy making and initiatives such as Erase Bullying or British Columbia Education Plan. Education.data.gov[63] provides a wealth of information about education in the USA. The Open Data inventory[64] provides more data reported to the Department of Education. In New Zealand open government data on schools in an app[65] to help you find schools in the local area.

Bahawalpur Service Delivery Unit (BSDU)[66], an initiative by the Government of Punjab province in Pakistan, aims to engage citizens in the governance of service delivery. Led by Imran Sikandar Baloch, District Coordination Officer of Bahawalpur district in Punjab, this initiative is built on open data and has already delivered increased attendance of teachers and students over the past year. Technology and design partner for this initiative is Technology for People Initiative based at the Lahore University of Management Sciences. It features a mobile app that allows officials and citizens to monitor attendance by teachers and students at school. The information is aggregated online and made publicly accessible. The aim is to enable and motivate citizens to collect, analyze and disseminate service delivery data in order to drive performance and help effective decision making. The initiative has led to improved teacher attendance, which in turn has led to improved pupil grades. By showing how open data can help in the developing world, BDSU won the Making Voices Count global innovation competition.

Check My School is a social accountability initiative designed and instituted by the Affiliated Network for Social Accountability in East Asia and the Pacific (ANSA-EAP), and uses a blended approach through on the ground mobilization effort and community monitoring, tapping modern technology as a key tool. The CMS project is supported by the Open Society Institute and the World Bank Institute[67].

Other activities worth nothing are Education GPS[68], the OECD source for internationally comparable data on education policies and practices, opportunities and outcomes. Accessible any time, in real time, the Education GPS provides the latest information on how countries are working to develop high-quality and

---

[62] http://www.data.gov.bc.ca/.

[63] http://www.data.gov/education/.

[64] http://datainventory.ed.gov/AboutTheInventory.

[65] https://itunes.apple.com/us/app/recredible/id620437846?mt=8.

[66] http://ideas.makingallvoicescount.org/a/dtd/Bahawalpur-Service-Delivery-Unit-BSDU/18743-26650.

[67] http://www.checkmyschool.org.

[68] http://gpseducation.oecd.org.

equitable education systems. The Pearson Learning Curve Index[69] combines national data and a number of international rankings – including PISA, TIMSS and PIRLS – to provide an interpretation of how countries systems are performing relative to each other.

Higher Education. In the UK the Open Data Challenge identified applications of open education datasets in services like Which? University[70] which builds on the National Student Survey (NSS) annual survey held in Unistats, the Key information sets and other related datasets to allow aid students to select a university In Higher Education the development of equipment.data[71] has been funded by EPSRC in response to the need to improve visibility and utilisation of UK research equipment. This relatively simple technology enables searching across all published UK research equipment databases through one aggregation "portal", allowing greater accessibility with the aim to improve efficiency and stimulate greater collaboration in the sector. The data used is available to download from the site.

A more recent activity has seen Universities UK, a membership organisation for university leaders, run seminar series entitled Creating value from open data. The series has now led into a Jisc funded project with partners from Universities UK, the Open Data Institute, the National Union of Students and the Leadership Foundation. The project has 8 universities signed up: Edinburgh, Oxford, Cambridge, Newcastle, Aberdeen, The Open University, Southampton, Greenwich. Together they will work to develop a web based application and strive towards release of appropriate datasets release. The web app will focus on student recruitment, business processes, research management and the Research Excellence Framework, student experience (e.g., use of labour market information). The project will also include a data capability study and an exploratory look at the data skills curriculum[72].

The School of Data, through their data expeditions[73], are starting to do some important work in the area of education data in the developing world. And in January the World Bank released a new open data tool called SABER (The Systems Approach for Better Education Results), which enables comparison of countries education policies. The web tool helps countries collect and analyze information on their education policies, benchmark themselves against other countries, and prioritize areas for reform, with the goal of ensuring that in those countries all children and youth go to school and learn.

All over the world prototypes and apps are been developed that use and build on open education data.

---

[69] http://thelearningcurve.pearson.com/.

[70] http://university.which.co.uk.

[71] http://equipment.data.ac.uk/.

[72] See http://theodi.org/news/odi-uuk-and-top-uk-universities-launch-project–to-unlock-open-data-potential-in-higher-education.

[73] See http://education.okfn.org/school-of-data-using-education-data/.

There are still challenges that those keen to develop applications using open education data face. Privacy and data protection laws can often prevent access to some potentially useful datasets, yet many datasets that are not personal or controversial remain unavailable, or only available under a closed licence or inappropriate format. This may be for many reasons: trust, concerns around quality and cost being the biggest issues. Naturally there is a cost to releasing data but in many cases this can be far out-weighed by cost-savings later down the line, so for example a proactive approach is likely to save time and effort should Freedom of Information (FOI) requests be made.

Well-defined use cases are starting to emerge but can be still hard to find. The EU funded Open Discovery Space (ODS)[74] project aims to create a platform for teachers across Europe for sharing and repurposing of open educational resources.

However, ODS, also deals with mining data and usage for further improving the value chain of educational resources and open education. It creates a social data layer around education resources that crowd sources appreciation and usage data. Social data in this context is appreciation metadata that further describes a resource. It comprises intentional user inputs such as likert scale star ratings, comments, free or guided tags, shares, etc. From these datasets aggregations can be used in an infinite number of mashups to provide e.g. resource recommendations or karma measures. In addition, ODS also uses tracking data (called paradata) which collects users' activities in the ODS portal (e.g. looking at a resource, downloading, etc.). This allows for other statistical analytics such as most looked at, or most downloaded resource. In more sophisticated ways it also permits to draw conclusions about the similarity of users that looked at or downloaded the same resources or that follow similar type users. Analogous methods are well known from social networks (Facebook: "friends you may know", Twitter: "people who you may want to follow"), sales sites (Amazon: "people who looked at this also looked at"), or review portals (Tripadvisor: "most popular or most highly rated hotel").

ODS goes beyond collecting data from users of the portal alone, but also harvests social data from other OER portals. This is to say that if a user star-rates a resource in a sister portal to ODS, this rating will enter the ODS ratings data through a data harvesting cycle. In this way, opinion mining is not restricted to a single portal alone and enhances the value of the resource descriptor no matter where the users tag it. Harvesting social metadata from other portals encounters no legal obstacles, even if this data is not linked open data, because: (1) it is anonymous data and cannot be connected to a user's identity, (2) there is no copyright associated with protecting user expressions like star ratings, bookmarks or keyword tags. This is because it does not constitute an act of (substantial) creativity on behalf of the author of such social metadata. ODS not only re-uses social data from associated repositories, it also aims at exposing its own data as open linked data to other third party service providers. It has

---

[74] http://www.opendiscoveryspace.eu/project.

to be said, though, that paradata (recording user activities in the portal) is not going to be exposed due to ethical and privacy reasons.

6.5 Open Data in Education and Learning Analytics

Online education is producing vast amounts of data about students. Much of these online courses are openly available and the data from them should be too. The data will enable academic institutions and course providers to deliver their courses more efficiently and more appropriately to their students. It will also allow students to personalize their educational experience to best suit their needs. Data collected can include administrative data, demographic information, grade information, attendance and activity data, engagement metrics, course selection etc.

Learning analytics is defined as the measurement, collection, analysis and reporting of data about learners and their contexts, for purposes of understanding and optimising learning and the environments in which it occurs.

Data from online courses can enable grade prediction and student success, measure student performance, improve student retention and determine what learners know and what they currently do not know. It can also monitor learner engagement, personalize learning which in turn can ensure relevant content is delivered. Other potential uses are the reduction of classroom administrative work. For further information on learning analytics see the Learning Analytics Community Exchange (LACE) Project[75] and the Society for Learning Analytics Research (SoLAR).[76]

6.6 Open Data in Education Challenges and Opportunities

The main benefits of using open data are around transparency, releasing social and commercial value, and participation and engagement. By opening up data, citizens are enabled to be much more directly informed and involved in decision-making. Open education data holds huge potential for students, schools and institutions and governments and policy makers. However there are challenges that need to be addressed.

One clear double edged sword relates to the monetary value of data. The Omidyar Network believe open data, including open access research, could contribute as much as $13tn to the economies of the G20 nations cumulatively over the next five years. This contribution is primarily in transparency and improved efficiencies. The Open Data Institute (ODI), a private limited company established as a not-for-profit organisation set up by the UK government to catalyse the evolution of an open data culture to create economic, environmental, and social value is a good example here. The ODI aims to unlock supply, generate demand, create and disseminate knowledge to address local and global issues. They have carried out work looking at business models in the open data space.

---

[75] http://www.laceproject.eu/.

[76] http://solaresearch.org/.

Their guide How to make a business case for open data[77] offers three general business models:

- freemium: you provide an "added value" data product or service, for which you charge
- cross subsidy: you reach more customers, or provide enhanced services to existing customers, through wider sharing and use of your data
- network effects: by collaborating with other organisations, you reduce your costs in maintaining data which you use in your work or extend the possible audience for your products and services

However monetary gain also has issues and trading in data has shown a clear upward trend. Discussions in a recent Twitter chat facilitated by Marketplace[78] noted that "Schools are often handing over datasets to private industry with few checks and balances on whether it can be warehoused, bought, sold, and traded – all to develop product to sell back at great expense to public education. The key question is the product really furthering public education or is it just lining the pockets of the rich?

Opening anything up makes organisations more vulnerable, especially if they have something to hide, or if their data is inaccurate or incomplete. There is also a cost to releasing and building on data. Often this cost is outweighed by the social or economic benefit generated, but this benefit can develop over time so can be hard to demonstrate.

Other matters of contention include the possible misinterpretion or misrepresention of data, privacy and ownership and the measuring and monitoring of individuals.

6.7   Open Education Handbook

The LinkedUp Project description of work initially described the 'LinkedUp Handbook on Open Data in Education' as a "resource for both educators and Web data providers as well as adopters... The LinkedUp Handbook will be created as a living document to reflect project learnings and findings, which will help others, both during the project and beyond it". To fulfill this brief over time the handbook[79] has evolved to consider the broader scope of open education resulting in it being renamed as 'the Open Education Handbook'. During its evolution the handbook has received contributions from organisations and individuals that span sectors and countries. The writing of the handbook has been very much embedded within the Open Education Working Group and will continue to remain an important part of working group work. Embedding the writing of handbook in such a group has ensured that it is part of a committed community made up of practitioners working in open education and those interested in its broader implications.

---

[77] See http://theodi.org/guides/how-make-business-case-open-data.

[78] See http://www.marketplace.org/topics/education/learningcurve/our-readers-worries-and-hopes-about-student-data.

[79] http://education.okfn.org/handbook/.

The handbook is living web document targeting educational practitioners and the education community at large and it been crowd-sourced and drafted over a series of online and offline events. The initial booksprint held in London and was attended by education experts from different sectors (commercial, academic, government, not-for-profit). A second booksprint took place in Berlin on Friday 22nd November 2013 and was organised in collaboration with Wikimedia Deutschland[80]. During this event the handbook was 'chunked up' into a number of question areas and discussion took place over the direction of the handbook. On January 20 th 2014, as an activity for Education Freedom Day, the Open Education Handbook was translated and adapted into Portuguese. This process highlighted some interesting possibilities and challenges for the handbook such as the requirements of a global audience[81]. A timeline event also took place at which a group physically mapped important open education events, which were then added to an online timemap.

Throughout 2014 the handbook has been further developed through a series of Friday Chats that have taken place on the Open Education Working Group[82]. These discussions have provided the handbook with well-thought out objective content that is not available elsewhere on the web. In late September 2014 in preparation for the delivery of the 'final version' of the handbook an external editor was employed to proof read the handbook. The editor was asked to look at areas including overall structure, typos and poor writing, universal style, fact checking, citations and links, glossary and definitions.

Content is key within the handbook and it has a broad coverage considering both practical and factual areas and more discursive topics. Some of the questions it intents to help answer are: What is open? What is education? What is open education? Is traditional education not open? What affect does open education have on education? Who is meant to benefit from open education? What are open educational resources (OERs)? What are Open Licences? What is Open Learning and Practice? What is Open Policy? What is open education data? How does open data relate to open education? When possible references are given to examples and related projects and initiatives. It includes a section on open education data that considers the drivers behind data release and use, available technologies, data use in the developing world, open data competitions and current case studies.

The handbook outline was created using three Google documents. In late 2013 the handbook was moved from Google Docs to Booktype[83], an open source platform for writing and publishing print and digital books developed by Source-Fabric. It has continued to be written in Booktype and the software has been found to be a suitable platform in which to house a collaboratively written

---

[80] See http://education.okfn.org/second-open-education-handbook-booksprint-berlin/.

[81] Manual de Educao Aberta – http://education.okfn.org/manual-de-educacao-aberta/.

[82] See https://pad.okfn.org/p/Open_Education_Working_Friday_Chats.

[83] https://www.sourcefabric.org/en/booktype/.

handbook. As a resource the handbook offers an introduction to various topic areas but is also a springboard from which users can connect with other relevant resources. These connections are achieved by links and so the handbook is by nature hyperlink heavy. While it is possible to create a downloadable version of the handbook it is clear that PDF or Word are not the optimum mediums in which to view it. Prior to the delivery date for deliverables LinkedUp requested that the handbook could be delivered in two versions: Firstly an online version that is optimised for those viewing on the web Secondly an open ebook format that can be viewed on a computer while online or offline, and can also be printed. These formats differ from the usual EU deliverable format but it is hoped that the end-result is a useable, user-friendly output that can be reused. The current version of the handbook is now available online, as an ePub book and as a PDF. The Handbook is available under a Creative Commons Attribution 4.0 International (CC BY 4.0)[84].

The handbook is now a comprehensive and intelligent overview of the current situation with regard to Open Education and Open Education data. Supporting a large and disparate community to collaboratively produce an open resource has posed many interesting challenges: for example how do you resolve differences of opinion? Is targeting for audiences possible? What processes need to be in place to verify content? To realise its full potential such a resource needs to be allowed to continue to evolve and be built upon. As explained previously, the writing of the handbook has been very much embedded within the Open Education Working Group throughout the LinkedUp Project lifecycle, and it is here that it will continue to stay until a more appropriate place is found. Discussions have already taken place around the future of the handbook and possible ideas include moving it to Wiki books, embedding it within Wikipedia and building a front-end for it to use with Booktype. It is hoped that these ideas can be developed further in discussion with the community. An online community session is planned for early December that will explore the future development of the handbook and appropriate delivery mechanisms.

7   Connections with Other Groups

The Open Education Working Group is not working alone in the open education and open data space. Although it may be the only group to span the data and resources space there are many other more focused groups with which it has a lot in common. It will continue to connect with these groups and support their mission statements. Many of these groups are identified in the Open Education Handbook[85].

A formal connection has been made with the W3C Open Linked Education community group, a focus point for the community to collect, capture and adopt the practices that are going to be the foundation of the web of educational data.

---

[84] https://creativecommons.org/licenses/by/4.0/.

[85] See http://booktype.okfn.org/open-education-handbook-2014/oer-communities-and-interest-groups/.

The group brings together existing to gather initiatives the practices currently employed to sharing education-related data on the web including vocabularies and best practices. The LinkedUp consortium will lead the community group from autumn 2014.

8   Conclusions

The Open Education Working Group is an engaged, multi-faceted, global community with broad interests related to open data in education. Through online and offline events it has supported a two-way process in which interesting and valuable dialogues have taken place. The long-term impact of the Open Education Working Group and its new and growing community is difficult to measure at this stage. However, issues around data creation and use in education (such as privacy, measurement of learning, online learning, data-driven decision making) are only likely to come closer to the fore over the next few years. Currently the US and European countries are leading the way in exploration of the potential of open data in education, but the next few years may well see other countries, such as Brazil, or those in the global south, picking up the baton. The Open Education Working Group want to support these groups around the world interested in applying the power of open data in education and using it for social good. Its intention is to bring many more people into the conversations that are starting to take place. The challenges at this stage are to ensure the sustainability of the group and that it strives for global coverage rather than the current US, UK and Euro-centric focus. With a growing, vibrant community it is a challenge it is well-equipped to take on.

References

1. Daga, E., d'Aquin, M., Adamou, A., Brown, S.: The open university linked data - data.open.ac.uk. Seman. Web J. (2016) (to appear)
2. Dietze, S., Taibi, D., d'Aquin, M.: Facilitating scientometrics in learning analytics and educational data mining - the lak dataset. Seman. Web J. (2016) (to appear)
3. Guy, M., d'Aquin, M., Dietze, S., Drachsler, H., Herder, E., Parodi, E.: Linkedup: Linking open data for education. Ariadne **72**, 70–74 (2014)
4. Reinikka, R., Svensson, J.: Local capture: Evidence from a central government transfer program in uganda. Q. J. Econ. **2**(119), 679–705 (2004)
5. World Economic Forum. Education, skills 2.0: New targets and innovative approaches. http://www.weforum.org/reports/education-and-skills-20-new-targets-and-innovative-approaches

Author Index

Printed in the United States
By Bookmasters